my revisi

AQA A-level

CHEMISTRY

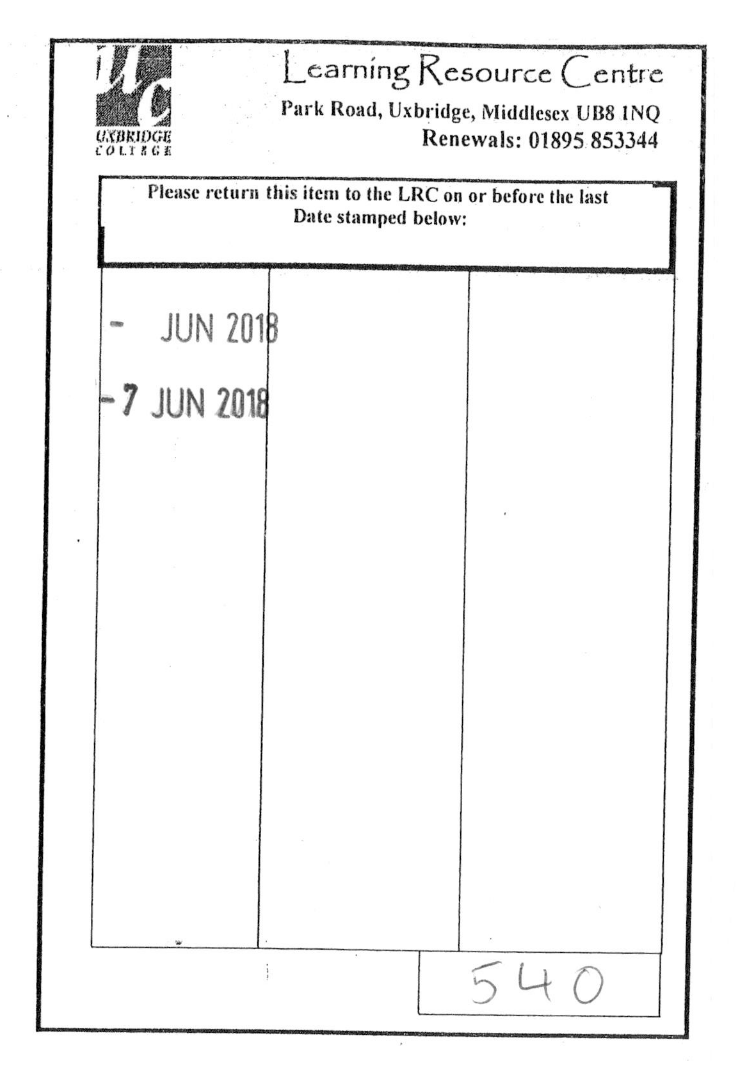

Rob King

HODDER EDUCATION

AN HACHETTE UK COMPANY

Hachette UK's policy is to use papers that are natural, renewable and recyclable products and made from wood grown in sustainable forests. The logging and manufacturing processes are expected to conform to the environmental regulations of the country of origin.

Orders: please contact Bookpoint Ltd, 130 Park Drive, Milton Park, Abingdon, Oxon OX14 4SE. Telephone: (44) 01235 827720. Fax: (44) 01235 400454. Email education@bookpoint.co.uk Lines are open from 9 a.m. to 5 p.m., Monday to Saturday, with a 24-hour message answering service. You can also order through our website: www.hoddereducation.co.uk

ISBN: 978 1 4718 4222 1

First published in 2016 by
Hodder Education,
An Hachette UK Company
Carmelite House
50 Victoria Embankment
London EC4Y 0DZ

www.hoddereducation.co.uk

Impression number 10 9 8 7 6 5 4 3 2 1
Year 2020 2019 2018 2017 2016

Cover photo reproduced by permission of Jag_cz/Fotolia

Typeset in Bembo Std Regular, 11/13 pts. by Aptara, Inc.

Printed in Spain

A catalogue record for this title is available from the British Library.

Get the most from this book

Everyone has to decide his or her own revision strategy, but it is essential to review your work, learn it and test your understanding. *My Revision Notes* will help you to do that in a planned way, topic by topic. Use this book as the cornerstone of your revision and don't hesitate to write in it — personalise your notes and check your progress by ticking off each section as you revise.

Tick to track your progress 

Use the revision planner on pages 4–7 to plan your revision, topic by topic. Tick each box when you have:

- revised and understood a topic
- tested yourself
- practised the exam questions and gone online to check your answers and complete the quick quizzes

You can also keep track of your revision by ticking off each topic heading in the book. You may find it helpful to add your own notes as you work through each topic.

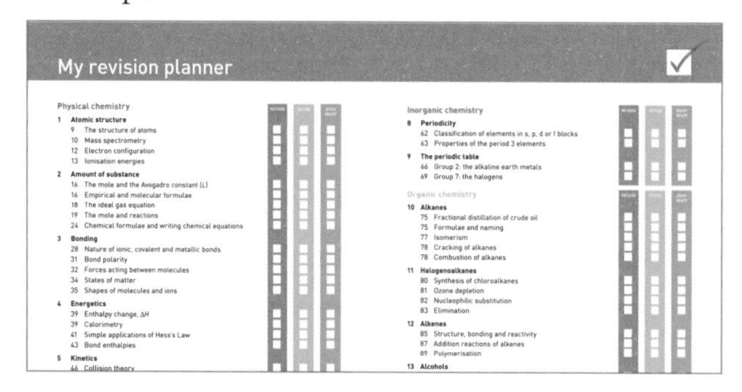
My revision planner

Physical chemistry

1 **Atomic structure**
- 9 The structure of atoms
- 10 Mass spectrometry
- 12 Electron configuration
- 13 Ionisation energies

2 **Amount of substance**
- 16 The mole and the Avogadro constant (*L*)
- 16 Empirical and molecular formulae
- 18 The ideal gas equation
- 19 The mole and reactions
- 24 Chemical formulae and writing chemical equations

3 **Bonding**
- 28 Nature of ionic, covalent and metallic bonds
- 31 Bond polarity
- 32 Forces acting between molecules
- 34 States of matter
- 35 Shapes of molecules and ions

4 **Energetics**
- 39 Enthalpy change, ΔH
- 39 Calorimetry
- 41 Simple applications of Hess's Law
- 43 Bond enthalpies

5 **Kinetics**
- 44 Collision theory

Inorganic chemistry

8 **Periodicity**
- 62 Classification of elements in s, p, d or f blocks
- 63 Properties of the period 3 elements

9 **The periodic table**
- 66 Group 2: the alkaline earth metals
- 69 Group 7: the halogens

Organic chemistry

10 **Alkanes**
- 75 Fractional distillation of crude oil
- 75 Formulae and naming
- 77 Isomerism
- 78 Cracking of alkanes
- 78 Combustion of alkanes

11 **Halogenoalkanes**
- 80 Synthesis of chloroalkanes
- 81 Ozone depletion
- 82 Nucleophilic substitution
- 83 Elimination

12 **Alkenes**
- 85 Structure, bonding and reactivity
- 87 Addition reactions of alkenes
- 89 Polymerisation

13 **Alcohols**

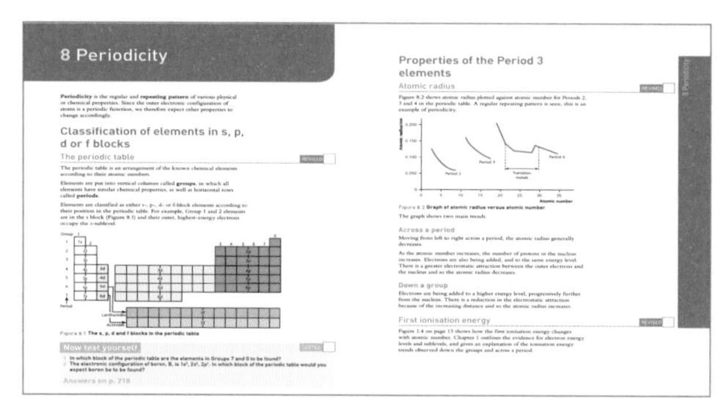
8 Periodicity

Classification of elements in s, p, d or f blocks

The periodic table

Now test yourself

Properties of the Period 3 elements

Atomic radius

Across a period

Down a group

First ionisation energy

Features to help you succeed

Exam tips and summaries

Throughout the book there are tips from examiners to help you boost your final grade. Summaries provide advice on how to approach each topic in the exams, and suggest other things you might want to mention to gain those valuable extra marks.

Typical mistakes

Examiners identify the typical mistakes students make and explain how you can avoid them.

Now test yourself

These short, knowledge-based questions provide the first step in testing your learning. Answers are at the back of the book.

Definitions and key words

Clear, concise definitions of essential key terms are provided where they first appear.

Key words from the specification are highlighted in bold throughout the book.

Revision activities

These activities will help you to understand each topic in an interactive way.

Exam practice

Practice exam questions are provided for each topic. Use them to consolidate your revision and practise your exam skills.

Online

Go online to check your answers to the exam questions and try out the extra quick quizzes at **www.hoddereducation.co.uk/myrevisionnotes**

My revision planner

Physical chemistry

Organic chemistry (A-level only)

REVISED | TESTED | EXAM READY

Exam practice answers and quick quizzes at www.hoddereducation.co.uk/myrevisionnotes

Countdown to my exams

6–8 weeks to go

- Start by looking at the specification — make sure you know exactly what material you need to revise and the style of the examination. Use the revision planner on pages 4–7 to familiarise yourself with the topics.
- Organise your notes, making sure you have covered everything on the specification. The revision planner will help you to group your notes into topics.
- Work out a realistic revision plan that will allow you time for relaxation. Set aside days and times for all the subjects that you need to study, and stick to your timetable.
- Set yourself sensible targets. Break your revision down into focused sessions of around 40 minutes, divided by breaks. These *Revision Notes* organise the basic facts into short, memorable sections to make revising easier.

REVISED ☐

2–5 weeks to go

- Read through the relevant sections of this book and refer to the exam tips, summaries, typical mistakes and key terms. Tick off the topics as you feel confident about them. Highlight those topics you find difficult and look at them again in detail.
- Test your understanding of each topic by working through the 'Now test yourself' questions in the book. Look up the answers at the back of the book.
- Make a note of any problem areas as you revise, and ask your teacher to go over these in class.
- Look at past papers. They are one of the best ways to revise and practise your exam skills. Write or prepare planned answers to the exam practice questions provided in this book. Check your answers online and try out the extra quick quizzes at **www.hoddereducation.co.uk/myrevisionnotes**
- Use the revision activities to try out different revision methods. For example, you can make notes using mind maps, spider diagrams or flash cards.
- Track your progress using the revision planner and give yourself a reward when you have achieved your target.

REVISED ☐

One week to go

- Try to fit in at least one more timed practice of an entire past paper and seek feedback from your teacher, comparing your work closely with the mark scheme.
- Check the revision planner to make sure you haven't missed out any topics. Brush up on any areas of difficulty by talking them over with a friend or getting help from your teacher.
- Attend any revision classes put on by your teacher. Remember, he or she is an expert at preparing people for examinations.

REVISED ☐

The day before the examination

- Flick through these *Revision Notes* for useful reminders, for example the exam tips, summaries, typical mistakes and key terms.
- Check the time and place of your examination.
- Make sure you have everything you need — extra pens and pencils, tissues, a watch, bottled water, sweets.
- Allow some time to relax and have an early night to ensure you are fresh and alert for the examination.

REVISED ☐

My exams

A-level Chemistry Paper 1

Date:..

Time: ....

Location: ....

A-level Chemistry Paper 2

Date:..

Time: ....

Location: ....

A-level Chemistry Paper 3

Date:..

Time: ....

Location: ....

1 Atomic structure

The structure of atoms

Protons, neutrons and electrons

REVISED

Table 1.1 Fundamental particles

Particle	Relative charge	Relative mass
Proton	+1	1
Neutron	0	1
Electron	−1	$\frac{1}{1836}$

The **atomic number** (Z) and **mass number** (A) can be used to deduce the number of protons, neutrons and electrons in any atom or ion.

To calculate the number of neutrons in the nucleus, the atomic number (equal to the number of protons) is subtracted from the mass number (equal to the number of protons + neutrons).

For example, the elements of calcium, potassium and phosphorus are represented as:

$^{40}_{20}Ca$ $^{39}_{19}K$ $^{31}_{15}P$

Atoms do not have an overall charge because the number of positively charged protons is the same as the number of negatively charged electrons. However, when an **ion** forms — either by losing or by gaining electrons — there is an overall charge because there will be unequal numbers of positive charges and negative charges.

Atomic number is the number of protons in the nucleus of an atom (or ion).

Mass number is the total number of protons plus neutrons in the nucleus.

Exam tip

Remember that when atoms lose electrons they form positively charged ions; when atoms gain electrons they form negatively charged ions.

Isotopes

REVISED

Many elements exist as **isotopes**. In a sample of chlorine, for example, 75% of the chlorine atoms have a mass number of 35 and 25% have a mass number of 37. These atoms have different masses because they have different numbers of neutrons.

In Table 1.2, the numbers of protons, neutrons and electrons are indicated for all three isotopes of magnesium, thus showing the differing numbers of neutrons.

Isotopes are atoms with the same number of protons, but different numbers of neutrons.

Table 1.2 The isotopes of magnesium (atomic number = 12)

Isotopes	Protons	Neutrons	Electrons
^{24}Mg	12	12	12
^{25}Mg	12	13	12
^{26}Mg	12	14	12

Mass spectrometry

The mass spectrometer

REVISED

The **time of flight (TOF) mass spectrometer** (Figure 1.1) is a device that enables substances (elements or compounds) to be analysed by determining the masses of ions formed.

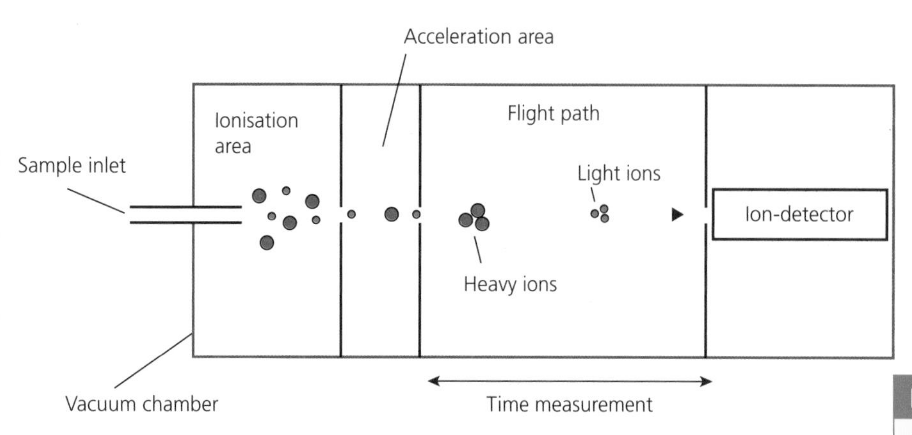

Figure 1.1 Diagram of a mass spectrometer

There are four stages involved in analysing a sample:

1. **Electron spray ionisation** — a high-energy beam of electrons from the electron gun removes the highest-energy (outer) electron from a molecule or an atom, forming a positively charged ion. For example, using gallium:

   $Ga(g) \rightarrow Ga^{+}(g) + e^{-}$

2. **Acceleration** — the positively charged ion is accelerated towards a negatively charged electrode. This provides all ions with the same kinetic energy.
3. **Ion drift** — the positively charged ions are electrostatically attracted to the negatively charged cathode.
4. **Ion detection and data analysis** — ions of different masses are detected electronically and produce a small, varying electrical current that can be amplified and displayed on a computer. A mass spectrum is produced.

Exam tip

Equations showing ionisation processes that take place in a mass spectrometer occur in the gas phase, so make sure that state symbols are added to equations showing ionisations.

Revision activity

Try to make up an acronym that will help you to remember the main processes involved when analysing a sample in a mass spectrometer.

How to interpret a mass spectrum

REVISED

The mass spectrum of lead is shown in Figure 1.2.

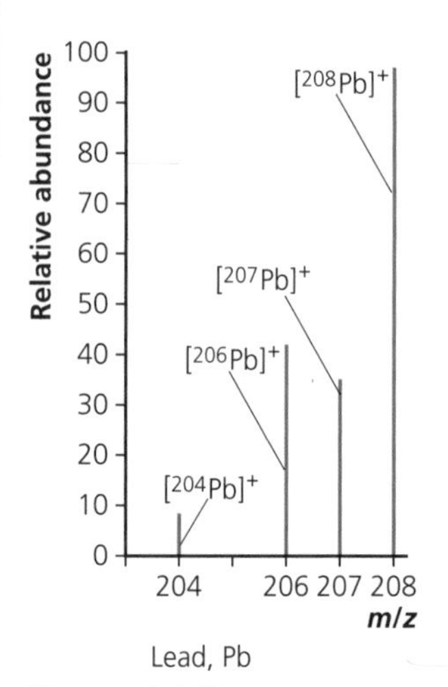

Figure 1.2 The mass spectrum of lead

- In the spectrum there are four peaks. This means that there are four isotopes of lead in the sample: ^{204}Pb, ^{206}Pb, ^{207}Pb and ^{208}Pb.
- There is more ^{208}Pb than any other lead isotopes because this is the tallest peak — this isotope gives the peak with the highest abundance.
- You can calculate a value for the **relative atomic mass** (A_r) of lead by multiplying each mass number by its abundance percentage (divided by 100) and then adding them all together. If the abundances in this case are ^{204}Pb (4.9%), ^{206}Pb (23.2%), ^{207}Pb (19.2%) and ^{208}Pb (52.7%), the calculation will be:

$$A_r \text{ for lead} = \left(\frac{4.9}{100} \times 204\right) + \left(\frac{23.2}{100} \times 206\right) + \left(\frac{19.2}{100} \times 207\right) + \left(\frac{52.7}{100} \times 208\right)$$

$$= 207.1 \text{ (no units)}$$

Mass spectrometers allow **relative molecular masses** and their abundances to be determined with a high degree of precision.

Mass spectrometers can be used in planetary space probes to analyse samples of material found on other planets, and this information can then be beamed to Earth. This way the elements can be determined, together with the identity of molecules.

Exam tip

In a mass spectrum, mass-to-charge ratio is measured on the horizontal axis and relative abundance is measured on the vertical axis. The charge of an ion is normally designed to be +1. In a few cases, a +2 ion may form when a second electron is removed.

Revision activity

Look up on the internet an element of your choice and find out about the relative abundances of its isotopes. Then work out the element's relative atomic mass to practice this type of calculation.

Now test yourself

TESTED

1 Deduce the number of protons, neutrons and electrons in each of:
(a) $^{19}_{9}F$
(b) $^{74}_{34}Se$
(c) $^{48}_{22}Ti^{2+}$
(d) $^{79}_{35}Br^-$
2 Calculate the relative atomic mass for a sample of krypton from the data in Table 1.3. Give your answer to 2 decimal places.

Table 1.3

Isotopes	$^{78}_{36}Kr$	$^{80}_{36}Kr$	$^{82}_{36}Kr$	$^{83}_{36}Kr$	$^{84}_{36}Kr$	$^{86}_{36}Kr$
% abundance	0.35	2.3	11.6	11.5	56.9	17.4

3 A sample of boron is found to have a relative atomic mass of 10.8. Assuming that there are only two isotopes of this element, ^{10}B and ^{11}B, determine the percentage of each isotope in the sample.

Answers on p. 215

Relative atomic mass (A_r) is the weighted mean mass of an atom compared with $\frac{1}{12}$ the mass of an atom of ^{12}C. ^{12}C has the value of 12.0000 on this scale.

Relative molecular mass (M_r) is the weighted mean mass of a molecule compared with $\frac{1}{12}$ the mass of an atom of ^{12}C. It is the sum of the relative atomic masses of the atoms in a molecule.

Electron configuration

Electrons occupy **energy levels** (or shells) when orbiting an atomic nucleus. Energy levels are made up of **sublevels**, or subshells.

Sublevels fill up in the order shown in Figure 1.3, lowest energy first.

> **Exam tip**
>
> **Sublevels** are made up of **orbitals** and one orbital of any type can only hold up to two electrons.

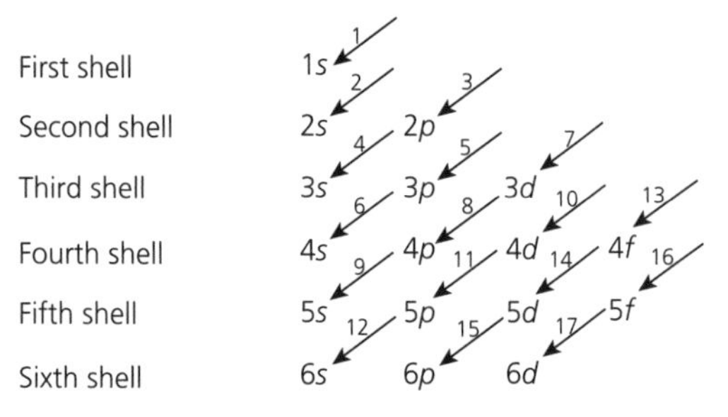

Figure 1.3 Arrangement of sublevels

You can write an electronic configuration for an atom or an ion by putting the correct number of electrons into each sublevel: 2 electrons in an *s*-sublevel; 6 electrons in a *p*-sublevel; 10 electrons in a *d*-sublevel and 14 in an *f*-sublevel. The number of electrons in a shell is given by $2n^2$, where n is the number of the shell. For example, in the third shell $n = 3$, so the number of electrons is 2×3^2, which is 18. Table 1.4 gives some examples.

Table 1.4 Some electron configurations

Element	Symbol/ion	Atomic number	Electronic configuration
Sodium	Na	11	$1s^2, 2s^2, 2p^6, 3s^1$
	Na^+		$1s^2, 2s^2, 2p^6$
Potassium	K	19	$1s^2, 2s^2, 2p^6, 3s^2, 3p^6, 4s^1$
	K^+		$1s^2, 2s^2, 2p^6, 3s^2, 3p^6$
Sulfur	S	16	$1s^2, 2s^2, 2p^6, 3s^2, 3p^4$
	S^{2-}		$1s^2, 2s^2, 2p^6, 3s^2, 3p^6$
Gallium	Ga	31	$1s^2, 2s^2, 2p^6, 3s^2, 3p^6, 4s^2, 3d^{10}, 4p^1$
	Ga^{3+}		$1s^2, 2s^2, 2p^6, 3s^2, 3p^6, 3d^{10}$
Krypton	Kr	36	$1s^2, 2s^2, 2p^6, 3s^2, 3p^6, 4s^2, 3d^{10}, 4p^6$

> **Exam tip**
>
> Notice that the 4*s*-sublevel fills up before the 3*d*-sublevel; this means that the 4th energy level starts to fill up before the 3rd energy level has been filled completely.

Now test yourself

TESTED

4 (a) Beryllium ($Z = 4$), magnesium ($Z = 12$) and calcium ($Z = 20$) are the first three members of Group 2 of the periodic table. Write out each element's electronic configuration.

(b) An ion of charge +3 has the electronic configuration $1s^2, 2s^2, 2p^6$. What is the atomic number, Z, for the ion?

Answers on p. 215

Ionisation energies

First ionisation energy

REVISED

The equation to represent the **first ionisation energy** of calcium is:

$Ca(g) \rightarrow Ca^{+}(g) + e^{-}$

A plot of first ionisation energy against atomic number is shown in Figure 1.4. This graph can provide valuable evidence for the electron arrangement in levels.

First ionisation energy is defined as the energy required to remove 1 mol of electrons from 1 mol of gaseous atoms under standard conditions (298 K and a pressure of 100 kPa).

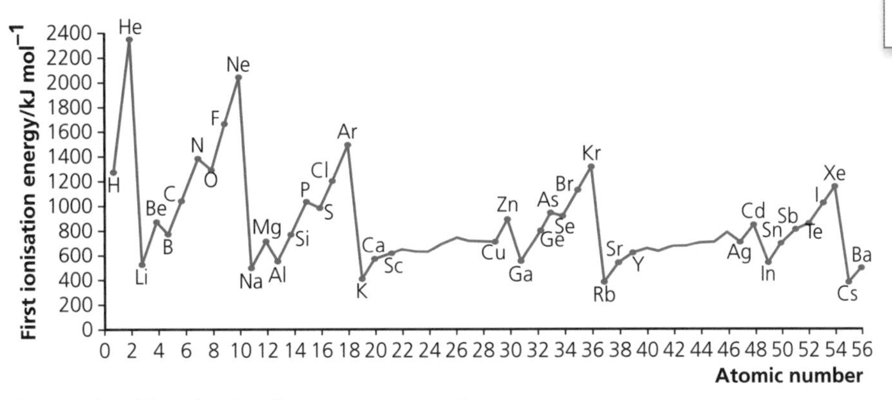

Figure 1.4 First ionisation energy graph

Energy levels

REVISED

It is possible to remove more than one electron from an atom if the energy of the bombarding electrons is sufficient. For example, the third ionisation energy of argon is represented as:

$Ar^{2+}(g) \rightarrow Ar^{3+}(g) + e^{-}$

Successive ionisation energies for elements give valuable evidence about the existence of energy levels.

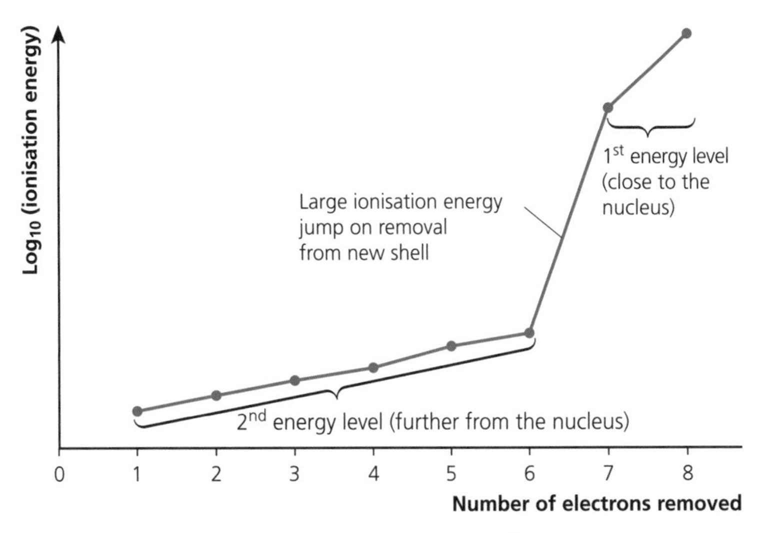

Figure 1.5 Graph of $\log_{10}$(ionisation energy) versus number of the electron being removed for oxygen

From Figure 1.5 it can be seen that:

- there is a large jump in ionisation energy required moving from the sixth to the seventh electron removal
- six electrons are relatively easy to remove from the nucleus, whereas two electrons are a lot more difficult to remove

The explanation is that the eight electrons must exist in two separate energy levels — two electrons are in an energy level closest to the positive charge of the nucleus and are therefore harder to remove. The other six electrons are in a higher energy level further from the nucleus and are therefore easier to remove.

Now test yourself

TESTED

5 (a) Sketch a graph to show the successive ionisation energies for silicon, $Z = 14$.
(b) Write an equation to show the fourth ionisation energy of phosphorus, P.

Answer on p. 215

Evidence for energy levels

Variation of first ionisation energies down Group 2 (Be–Ba)

The trend in first ionisation energies is a decrease in the energy on moving down the group.

The explanation is as follows:

- Despite the increased nuclear charge, the electron being removed is in a new energy level which is progressively further from the nucleus.
- The extra energy level provides extra shielding for the removed electron from the attraction of the positively charged nucleus.
- The net effect is to decrease the ionisation energy.

Evidence for energy sublevels

Variation of first ionisation energies across Period 3 from left to right

The trend in first ionisation energies is a general increase across Period 3 from sodium to argon.

The explanation is as follows:

- The electrons are being removed from the same electron energy level.
- The nuclear charge is increasing as more protons are being added from left to right.
- The electrons experience a greater attraction as the atom increases in atomic number.

Although the general trend is for ionisation energy to increase from left to right across a period, there are two small decreases.

From Group 2 to Group 3: magnesium to aluminium

The explanation for the decrease is as follows:

- Despite the increased nuclear charge, the added electron is in a new *p*-sublevel of slightly higher energy, and this is slightly further from the nucleus.
- The s^2 electrons, for the Group 3 element, provide some shielding.
- The overall effect is for the ionisation energy to decrease.

From Group 5 to Group 6, phosphorus to sulfur

The explanation for the decrease is as follows:

- Despite the increased nuclear charge, the electron from the Group 6 element is being removed from a p^4 configuration (Figure 1.6).

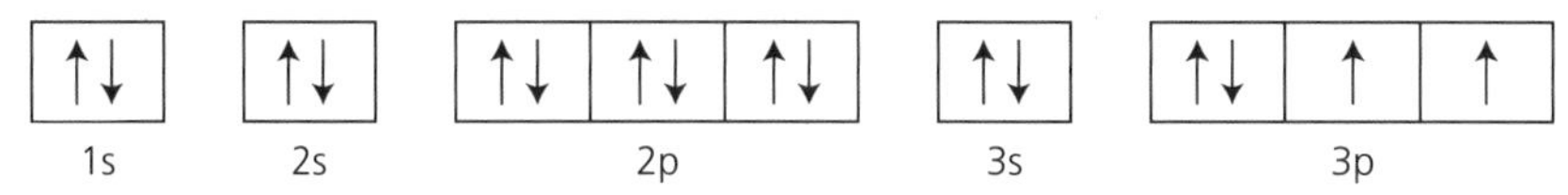

Figure 1.6 Arrangement of electrons in orbitals for sulfur

- There are four electrons in the outer a *p*-sublevel, so two of these must be paired in one orbital.
- This electron–electron repulsion lowers the attraction between the fourth electron and the nucleus.
- So the electron is easier to remove.

Exam practice

1 (a) Write the electron configuration of an Al^+ ion. [1]

(b) (i) State the meaning of the term 'first ionisation energy'. [2]

(ii) Write an equation, including state symbols, to show the reaction that describes the second ionisation energy of aluminium. [2]

(iii) Explain why the second ionisation energy of aluminium is higher than the first ionisation energy of aluminium. [2]

(c) State and explain the general trend in the first ionisation energies of the Period 3 elements sodium to chlorine. [3]

(d) Explain why sulfur has a lower first ionisation energy than phosphorus. [2]

(e) Explain why argon (Z = 18) has a much higher first ionisation energy than potassium (Z = 19) even though potassium has a larger positive nuclear charge. [2]

2 A sample of sulfur consisting of three isotopes has a relative atomic mass of 32.16. Table 1.5 gives the relative abundances of two of these isotopes.

Table 1.5

Mass number of isotope	32	33
Relative abundance/%	91.0	1.8

Use this information to determine the relative abundance and hence the mass number of the third isotope. Give your answer to the appropriate number of significant figures. [4]

3 Which of these atoms has the smallest number of neutrons? [1]

A ^{3}H
B ^{4}He
C ^{5}He
D ^{4}Li

Answers and quick quiz 1 online

ONLINE

Summary

You should now have an understanding of:

- the properties of protons, neutrons and electrons
- isotopes
- how a time of flight mass spectrometer works
- how a mass spectrum can be interpreted to provide information about isotopes
- how to calculate the relative atomic mass of an element using its mass spectrum
- electronic arrangement in terms of *s*, *p*, *d* notation
- ionisation energy
- the evidence for energy levels and sublevels from ionisation energies

2 Amount of substance

The mole and the Avogadro constant (*L*)

Definitions

REVISED

The mass of **1 mole** of an element or compound is its relative atomic mass (A_r) or relative molecular mass (M_r) expressed in grams.

An **amount of substance** is measured in moles:

$$\text{number of mol} = \frac{\text{mass of substance in g}}{\text{mass of 1 mol of that substance}}$$

or:

mass of substance = mass of 1 mol × number of moles

Example

You are provided with a sample of 0.56 mol of potassium dichromate(VI) ($K_2Cr_2O_7$). Calculate:

(a) the mass of the sample
(b) the number of potassium ions that would be present

Answer

(a) Mass of 1 mol of $K_2Cr_2O_7$ = (2 × 39.0) + (2 × 52.0) + (7 × 16.0) = 294 g
mass of substance = mass of 1 mol × number of mol
0.56 mol will have a mass of 294 g × 0.56 = 164.64 g

(b) For every 1 mol of potassium dichromate(VI) ($K_2Cr_2O_7$) there are 2 mol of potassium ions. Therefore, 0.56 mol of potassium dichromate(VI) would contain 0.56 × 2 mol of potassium ions, that is 1.12 mol.
The number of particles in 1 mol is given by the Avogadro constant, 6×10^{23}, so the number of potassium ions present will be $1.12 \times 6 \times 10^{23}$, that is 6.72×10^{23}.

Exam tip

Sometimes the term 'mol' is used to abbreviate the term 'moles', especially when it follows a number.

1 mole is defined as the amount of substance that contains as many elementary particles as there are atoms in exactly 12 g of the ^{12}C isotope.

The number 6×10^{23}, referred to as the **Avogadro constant**, is the number of specified particles — electrons, atoms, molecules or ions — in 1 mole.

Exam tip

The first step is always to work out the mass of 1 mol. This is then *multiplied* by the number of moles to calculate a mass.

Empirical and molecular formulae

Definitions

REVISED

The molecular formulae for ethane and hydrogen peroxide are C_2H_6 and H_2O_2 respectively, whereas the simplest whole-number ratios for these formulae — the empirical formula — are CH_3 and HO respectively.

An **empirical formula** is a formula representing the simplest whole-number ratio of atoms of each element in a compound.

A **molecular formula** represents the actual numbers of atoms of each element in one molecule.

Example 1

A hydrocarbon contains 2.51 g of carbon and 0.488 g of hydrogen. What is the empirical formula of the hydrocarbon?

Answer

mass of carbon = 2.51 g **mass of hydrogen = 0.488 g**

Convert into moles by dividing by the A_r for each element:

$$\text{mol of C} = \frac{2.51}{12.0} = 0.209\ \text{mol} \qquad \text{mol of H} = \frac{0.488}{1.0} = 0.488\ \text{mol}$$

Simplify the ratio by dividing each by the smaller number of moles:

$$\frac{0.209}{0.209} = 1 \qquad \frac{0.488}{0.209} = 2.33$$

Therefore, the ratio C : H is 1 : 2.33 or 3 : 7 (by multiplying each by 3 to get whole numbers). So the empirical formula is C_3H_7. If the relative molecular mass was given as 86 then the molecular formula would be C_6H_{14} since two 'lots' of the empirical formula would be required to give this molar mass.

Exam tip

Try to recognise certain ratios as being whole-number ratios in 'disguise', for example, 1 : 1.5 is 2 : 3; 1 : 1.33 is 3 : 4 and so on.

Questions featuring percentage compositions may also be set.

Example 2

A compound was found to contain 40.0% sulfur and 60.0% oxygen. What is the empirical formula for the compound? [A_r data: S = 32.1; O = 16.0]

Answer

If we assume the total mass of the compound is 100 g, then the masses of sulfur and oxygen will be 40.0 g and 60.0 g respectively.

mass of sulfur = 40.0 g **mass of oxygen = 60.0 g**

Convert into moles:

$$\frac{40.0}{32.1} = 1.25\ \text{moles} \qquad \frac{60.0}{16.0} = 3.75\ \text{moles}$$

The ratio of sulfur to oxygen is 1.25 : 3.75 or 1 : 3, so the empirical formula of the compound is SO_3.

Exam tip

Remember to divide each amount in moles by the smallest value — this will normally give new numbers that are easier to recognise as a whole-number ratio.

Now test yourself

TESTED

1. A compound is found to contain 1.00 g calcium and 1.77 g of chlorine only. What is its empirical formula? [A_r data: Ca = 40.1; Cl = 35.5]
2. A compound of calcium, silicon and oxygen is found to contain 0.210 g of calcium, 0.147 g of silicon and 0.252 g of oxygen. What is its empirical formula? [A_r data: Si = 28.1; Ca = 40.1; O = 16.0]
3. An oxide of nitrogen was found to contain 30.4% nitrogen by mass. The M_r of the oxide is 92.0. [A_r data: N = 14.0; O = 16.0]
   (a) What is the empirical formula of the oxide?
   (b) What is the molecular formula of the compound?
4. Caffeine is an organic molecule found in various natural products like coffee beans. Its displayed structure is shown in Figure 2.1.

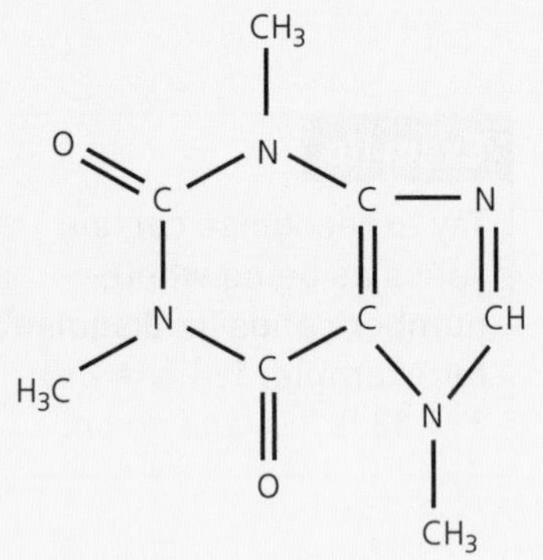

Figure 2.1

   (a) What is the molecular formula of a caffeine molecule?
   (b) A sample of coffee was analysed and found to contain 5.60×10^{-3} g of caffeine.
      (i) How many molecules of caffeine would this mass represent?
      (ii) The sample is then burned in excess oxygen. What is the maximum number of carbon dioxide molecules that could be produced in the combustion?
      [A_r data: N = 14.0; C = 12.0; O = 16.0; H = 1.0]

Answers on p. 215

The ideal gas equation

Calculations

REVISED

This states that:

$$pV = nRT$$

where p is the pressure measured in pascals (Pa), V is the volume measured in m^3, n is the number of moles, R is the gas constant ($8.31\,J\,K^{-1}\,mol^{-1}$) and T is the temperature in kelvin.

Exam tip

It is essential that the quantities in this equation have the correct units — notice in particular that volume of gas is in m^3, not cm^3 or dm^3.

Example

In an experiment, 0.700 mol of CO_2 was produced. This gas occupied a volume of 0.0450 m^3 at a pressure of 100 kPa. Calculate the temperature of the CO_2 and state the units of your answer.

Answer

$pV = nRT$

$p = 100000$ Pa $V = 0.0450$ m^3 $n = 0.700$ mol $R = 8.31$ J K^{-1} mol^{-1}

Substituting gives:

$100000 \times 0.0450 = 0.700 \times 8.31 \times T$

Rearranging gives:

$$T = \frac{100000 \times 0.0450}{0.700 \times 8.31} = 774\text{ K (to 3 significant figures)}$$

This temperature, in Celsius, would be 774 − 273 = 501°C.

Now test yourself

TESTED

5 A 0.905 mol sample of hydrogen gas, H_2, occupies a volume of 0.0330 m^3 at a temperature of 200°C. What is the pressure exerted by the gas? The gas constant, R is 8.31 J K^{-1} mol^{-1}.

Answers on p. 215

The mole and reactions

Questions on the mole are asked on many examination papers, so it is essential that this type of question is mastered.

Mass calculations

REVISED

Example

23.0 g of calcium carbonate decomposes fully on heating. Calculate the mass of carbon dioxide gas that forms. [A_r data: C = 12.0; Ca = 40.1; O = 16.0]

Answer

The equation for the reaction taking place is:

$CaCO_3(s) \rightarrow CaO(s) + CO_2(g)$

Step 1

Calculate the number of moles of calcium carbonate.

molar mass of $CaCO_3$ = 40.1 + 12.0 + (3 × 16.0) = 100.1 g

$$\text{number of mol of } CaCO_3 = \frac{\text{mass used (g)}}{\text{mass of 1 mol (g)}}$$

$$= \frac{23.0}{100.1} = 0.230\text{ mol}$$

Step 2

From the equation, $CaCO_3(s) \rightarrow CaO(s) + CO_2(g)$, 1 mol of $CaCO_3$ gives 1 mol of CO_2; the ratio is 1 : 1. So, the amount of CO_2 that forms is also 0.230 mol.

Step 3

Calculate the mass of carbon dioxide.

mass of CO_2 = mass of 1 mol of CO_2 × number of mol

= (12 + 16 + 16) × 0.230 = 10.1 g of CO_2 gas

Now test yourself

TESTED

6 Use the equation $Mg(s) + 2HCl(aq) \rightarrow MgCl_2(aq) + H_2(g)$ to determine the volume of hydrogen at 25°C that forms when 2.00 g of magnesium is added to excess hydrochloric acid.
[A_r data: Mg = 24.3; H = 1.0]

Answer on p. 215

Exam tip

Remember the three steps: **moles** (work out the moles); **ratio** (using the balanced symbol equation); calculate the **mass** of product (using moles × mass of 1 mol).

Gas calculations

REVISED

Equal volumes of gases contain equal numbers of particles, hence the amounts (in moles) will be the same.

Example

100 cm^3 of hydrogen is reacted with (a) chlorine and (b) oxygen in reactions that go to completion (no reactants are left). What volume of gas is formed in each reaction?

Answer

(a) $H_2(g) + Cl_2(g) \rightarrow 2HCl(g)$
From the equation, 1 volume of H_2 reacts with 1 volume of Cl_2 to form 2 volumes of HCl.
So, 100 cm^3 of H_2 reacts with 100 cm^3 of Cl_2 to form 200 cm^3 of HCl.

(b) $2H_2(g) + O_2(g) \rightarrow 2H_2O(g)$
2 volumes of hydrogen react with 1 volume of oxygen to form 2 volumes of water vapour.
So, 100 cm^3 of hydrogen reacts with 50 cm^3 of oxygen to form 100 cm^3 of water vapour.

Revision activity

Using the internet, find out about the famous French chemist and physicist Joseph Louis Gay-Lussac.

Now test yourself

TESTED

7 What volume of hydrogen will react with 150 cm^3 of nitrogen in the following reaction, assuming the reaction goes to completion?

$N_2(g) + 3H_2(g) \rightarrow 2NH_3(g)$

Answer on p. 215

Other reactions involving gases

REVISED

Example

Calculate the volume of gas produced at 298 K and 100 kPa when 1.45 g of lithium metal reacts with water. [A_r data: Li = 6.9]

Answer

$2Li(s) + 2H_2O(l) \rightarrow 2LiOH(aq) + H_2(g)$

amount of lithium $= \frac{1.45}{6.9} = 0.210$ mol

amount of hydrogen produced will be $\frac{0.210}{2}$ (according to the ratios in the equation)

So:

amount of hydrogen = 0.105 mol

volume of hydrogen $= \frac{nRT}{p} = \frac{0.105 \times 8.31 \times 298}{100\,000}$

$= 2.600 \times 10^{-3}\,m^3$ or $2600\,cm^3$

Revision activity

On a revision card, write formulae for how the amount of substance can be worked out for solids, solutions and gases. You will find these formulae very useful.

Now test yourself

TESTED

8 What volume of oxygen forms at room temperature and pressure when $100\,cm^3$ of a $0.500\,mol\,dm^{-3}$ solution of hydrogen peroxide decomposes according to this equation?

$2H_2O_2(aq) \rightarrow 2H_2O(l) + O_2(g)$

Answer on p. 215

Solutions

REVISED

The basic relationship for the amount of a substance in a solution can be expressed as:

$$\text{amount of solute dissolved (mol)} = \frac{\text{volume of solution (cm}^3) \times \text{concentration (mol dm}^{-3})}{1000\,\text{cm}^3}$$

It is also possible to convert everything into dm^3 and use:

$$\text{amount of solute dissolved (mol)} = \text{volume of solution (dm}^3) \times \text{concentration (mol dm}^{-3})$$

Example

Calculate the number of moles of acid dissolved in $25.50\,cm^3$ of $2.50 \times 10^{-3}\,mol\,dm^{-3}$ sulfuric acid.

Answer

number of mol $= \frac{25.50\,cm^3}{1000\,cm^3} \times 2.5 \times 10^{-3}\,mol\,dm^{-3}$

$= 6.375 \times 10^{-5}$ mol of H_2SO_4

Now test yourself

TESTED

9 How many moles of solute are dissolved in:
(a) 10.0 cm^3 of 0.200 mol dm^{-3} NaOH?
(b) 250 cm^3 of 1.20 mol dm^{-3} HNO_3?

Answer on p. 216

Solutions and reactions

Many reactions are carried out in solution. Calculations involving reacting amounts and volumes of solutions are common in examinations.

Exam tip

The stages involved in this type of calculation are the same as with mass calculations — work out the moles followed by the reaction ratio, and finally the volume (or concentration).

Example

Calculate the volume of 0.200 mol dm^{-3} sulfuric acid required to react exactly with 10.5 cm^3 of 0.400 mol dm^{-3} sodium hydroxide solution.

Answer

$$2NaOH(aq) + H_2SO_4(aq) \rightarrow Na_2SO_4(aq) + 2H_2O(l)$$

$$\text{number of moles of sodium hydroxide} = \frac{10.5}{1000} \times 0.400 = 4.2 \times 10^{-3}\ \text{mol}$$

According to the equation, 4.2×10^{-3} mol of sodium hydroxide reacts with ½(4.2×10^{-3}) moles of sulfuric acid (the reacting ratio according to the equation is 2 : 1). So, 2.1×10^{-3} moles of sulfuric acid are required. If the starting concentration is 0.2 mol dm^{-3}, then:

$$2.10 \times 10^{-3}\ \text{mol} = \frac{\text{volume in cm}^3}{1000\ \text{cm}^3} \times 0.200\ \text{mol dm}^{-3}$$

Rearranging, the volume is calculated as:

$$\frac{2.10 \times 10^{-3} \times 1000}{0.200} = 10.5\ \text{cm}^3$$

Revision activity

Write out the key stages involved in carrying out an acid–base titration. Find out why this method of working is considered to be very accurate.

Now test yourself

TESTED

10 Calculate the volume of 0.0500 mol dm^{-3} NaOH that will react exactly with 20.0 cm^3 of 0.900 mol dm^{-3} HCl.

Answer on p. 216

Percentage atom economy

REVISED

Sustainable development involves maximising our use of the resources available and reducing waste products if at all possible.

In a reaction, a measure of how much of the total mass of reactants is converted into the desired product is called the **atom economy** of that reaction, and is defined as:

$$\frac{\text{maximum mass of product}}{\text{total mass of all products}} \times 100$$

Or as:

$$\frac{\text{mass of product}}{\text{total } M_r \text{ values for all reactants}} \times 100$$

Exam tip

The symbols A_r and M_r mean *relative* atomic mass and *relative* molecular mass respectively. Neither term has any units as they are relative, or comparative.

Example

Calculate the atom economy for ethanol formation in this process:

$C_6H_{12}O_6(aq) \rightarrow 2C_2H_5OH(aq) + 2CO_2(g)$

Answer

M_r of glucose = 180, so the percentage atom economy will be:

$$\frac{2 \times 46}{180} \times 100 = 51.1\%$$

Now test yourself

TESTED

11 What is the atom economy for the formation of chloromethane in the reaction:

$CH_4(g) + Cl_2(g) \rightarrow CH_3Cl(g) + HCl(g)$

M_r for CH_3Cl is 50.5 and for HCl it is 36.5.

Answer on p. 216

Percentage yield

REVISED

In a reaction, the amount of product is called its **yield**:

$$\% \text{ yield} = \frac{\text{mass of product formed}}{\text{maximum theoretical mass of product}} \times 100$$

Example

When a 15.0 g sample of magnesium is heated in oxygen, it is found that 22.2 g of magnesium oxide forms. What is the percentage yield? [A_r data: Mg = 24.3; O = 16.0]

Answer

Write the chemical equation for the reaction:

$2Mg(s) + O_2(g) \rightarrow 2MgO(s)$

Calculate the actual mass of magnesium oxide expected:

$$\text{amount of magnesium} = \frac{15.0}{24.3} = 0.617\,\text{mol}$$

amount of MgO formed = 0.617 mol

mass of MgO = 0.617 × 40.3 = 24.9 g

The yield is therefore $\frac{22.2}{24.9} \times 100 = 89.2\%$ (to 3 significant figures).

Chemical formulae and writing chemical equations

Formulae of common ions

REVISED

The ability to write correct formulae is an important skill and should be practised. The same is true of chemical equations. Table 2.1 lists some formulae of common ions that are worth knowing.

Table 2.1 Some common ions

Formula of ion	Name of ion
CO_3^{2-}	Carbonate
SO_4^{2-}	Sulfate(VI)
NO_3^-	Nitrate(V)
NH_4^+	Ammonium
OH^-	Hydroxide
SO_3^{2-}	Sulfate(IV) or sulfite
NO_2^-	Nitrate(III) or nitrite
HCO_3^-	Hydrogencarbonate
SiO_3^{2-}	Silicate
ClO_3^-	Chlorate(V)
PO_4^{3-}	Phosphate(V)

Revision activity

On a revision card, write down the ions in this table and try to remember them. Keep looking at the card until you know them all. Alternatively, you could write the formulae of the ions into a simple database and use it to test yourself.

You can work out the formulae for some common compounds using a periodic table for reference.

Example

What is the formula for each of these compounds:

(a) magnesium oxide
(b) copper(I) sulfide
(c) manganese(II) nitrate
(d) ammonium sulfate?

Answer

(a) Magnesium oxide is made up of Mg^{2+} and O^{2-} ions. The charges will cancel so the formula is MgO.
(b) Copper(I) sulfide is made of Cu^+ and S^{2-} ions, swapping over the charges (or valencies) gives Cu_2S.
(c) Manganese(II) nitrate is made of Mn^{2+} and NO_3^- ions. Swapping over the numbers gives $Mn(NO_3)_2$. Remember to use a bracket when multiplying more than one element by a number.
(d) Ammonium sulfate is made of ammonium ions, NH_4^+, and sulfate ions, SO_4^{2-}. Swapping over the valencies gives $(NH_4)_2SO_4$.

Exam tip

When deducing the formula for a compound made of ions, simply swap over the charges, or valencies, in their lowest ratio.

Now test yourself

TESTED

12 What is the formula for each of these compounds?
(a) sodium fluoride
(b) potassium sulfate
(c) aluminium hydroxide?

Answers on p. 216

Table 2.2 lists some important substances have formulae that are easier to know, rather than work out.

Table 2.2

Substance	Formulae
Sulfuric acid	H_2SO_4
Nitric acid	HNO_3
Hydrochloric acid	HCl
Hydrogen peroxide	H_2O_2
Ammonia	NH_3
Alkanes	C_nH_{2n+2}

Writing balanced symbol equations

REVISED

Many reactants and products will be evident from the text of the examination question; others you will be expected to know. For example, the reactions of acids are very important — these are summarised in Figure 2.2.

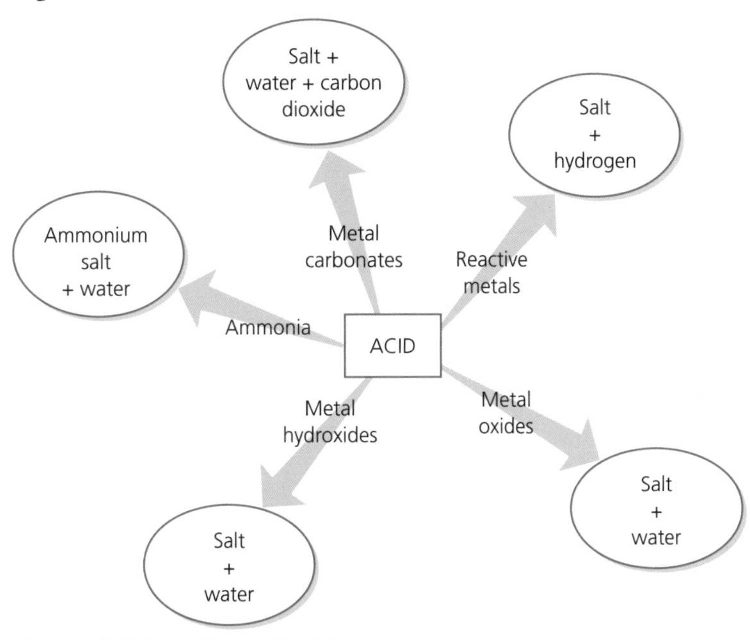

Figure 2.2 Reactions of acids

Example

Write a balanced equation to show how magnesium reacts with dilute nitric acid.

Answer

You need to know that magnesium nitrate and hydrogen gas form. You can then construct an equation and balance it.

- Write the formulae for each of the reactants and products:

  $Mg + HNO_3 \rightarrow Mg(NO_3)_2 + H_2$

- Balance to make it into an equation:

  $Mg + 2HNO_3 \rightarrow Mg(NO_3)_2 + H_2$

- Add state symbols if they are known:

  $Mg(s) + 2HNO_3(aq) \rightarrow Mg(NO_3)_2(aq) + H_2(g)$

Now test yourself

TESTED

13 Write a balanced symbol equation for the displacement reaction that takes place between aluminium metal and copper(II) sulfate.

Answers on p. 216

Ionic equations

REVISED

Ionic equations are useful for some reactions because they do not include spectator ions — just those ions that do take part in the reaction.

Consider the reaction mentioned previously:

$$Mg(s) + 2HNO_3(aq) \rightarrow Mg(NO_3)_2(aq) + H_2(g)$$

To convert this into an ionic equation, any substance that is dissolved in water has to be written out in its ionic form:

$$Mg(s) + 2H^+(aq) + 2NO_3^-(aq) \rightarrow Mg^{2+}(aq) + 2NO_3^-(aq) + H_2(g)$$

You then cancel out the spectator ions, in this case the nitrate ions, NO_3^-. This leaves:

$$Mg(s) + 2H^+(aq) \rightarrow Mg^{2+}(aq) + H_2(g)$$

which is the ionic equation for the reaction.

Exam tip

It is sometimes easier to remember the common types of ionic equation — neutralisation is $H^+(aq) + OH^-(aq) \rightarrow H_2O(l)$.

Now test yourself

TESTED

14 Write a balanced equation and an ionic equation to show potassium hydroxide solution reacting with hydrochloric acid. Include state symbols in your answer.

Answers on p. 216

Exam practice

1 Fluorine is a pale yellow gas that is known to be extremely reactive. It reacts with magnesium according to the equation:

$$Mg(s) + F_2(g) \rightarrow MgF_2(s)$$

In an experiment, 9.20 g of magnesium react with excess fluorine and magnesium fluoride is formed. [A_r data: Mg = 24.3; F = 19.0]

(a) Calculate the number of moles of magnesium used in the experiment. [1]
(b) Calculate the number of moles of magnesium fluoride formed in the reaction. [1]
(c) What mass of magnesium fluoride is formed? [1]
(d) What mass of magnesium would be needed if exactly 50.0 g of magnesium fluoride is required? [3]

2 In a titration, 25.00 cm^3 of 0.500 mol dm^{-3} sodium hydroxide is poured into to a conical flask and then titrated with hydrochloric acid of unknown concentration. It is found that 14.50 cm^3 of the acid is required for complete neutralisation. The equation for the reaction is:

$$HCl(aq) + NaOH(aq) \rightarrow NaCl(aq) + H_2O(l)$$

(a) Calculate the number of moles of sodium hydroxide used. [1]
(b) Deduce the number of moles of acid added from the burette. [1]
(c) Calculate the concentration of the dilute hydrochloric acid in mol dm^{-3}. [1]

3 The element indium forms a compound X with hydrogen and oxygen. Compound X contains 69.2% indium and 1.8% hydrogen by mass. Calculate the empirical formula of X. [A_r data: In = 114.8; H = 1.0; O = 16.0] [3]

4 An unknown metal carbonate reacts with hydrochloric acid according to the equation:

$M_2CO_3(aq) + 2HCl(aq) \rightarrow 2MCl(aq) + CO_2(g) + H_2O(l)$

A 1.72 g sample of M_2CO_3 was dissolved in distilled water to make 250 cm^3 of solution. A 25.0 cm^3 portion of this solution required 16.6 cm^3 of 0.150 mol dm^{-3} hydrochloric acid for complete reaction.

(a) Calculate the amount of HCl used. Give your answer to 3 significant figures. [1]
(b) Deduce the amount of M_2CO_3 that reacted with the hydrochloric acid. [1]
(c) How many moles of M_2CO_3 must have been present in the 250 cm^3 of solution? [1]
(d) What is the relative formula mass, M_r, of M_2CO_3? [1]
(e) Deduce the relative atomic mass of metal M. [1]

5 This question is about reactions of calcium compounds.
A pure solid is thought to be calcium hydroxide. The solid can be identified from its relative formula mass. The relative formula mass can be determined experimentally by reacting a measured mass of the pure solid with an excess of hydrochloric acid.
The equation for this reaction is $Ca(OH)_2 + 2HCl \rightarrow CaCl_2 + 2H_2O$.
The unreacted acid can then be determined by titration with a standard sodium hydroxide solution. You are provided with 50.0 cm^3 of 0.200 mol dm^{-3} hydrochloric acid.
Outline, giving brief practical details, how you would conduct an accurate experiment to calculate the relative formula mass of the solid using this method. [8]

6 Which of these pieces of apparatus has the lowest percentage error in the measurement shown? [1]
A Volume of 25 cm^3 measured with a burette with an error of ±0.1 cm^3.
B Volume of 25 cm^3 measured with a measuring cylinder with an error of ±0.5 cm^3.
C Mass of 0.150 g measured with a balance with an error of ±0.001 g.
D Temperature change of 23.2°C measured with a thermometer with an error of ±0.1°C.

7 A student is provided with a 5.00 cm^3 sample of 1.00×10^{-2} mol dm^{-3} hydrochloric acid.
The student is asked to devise a method to prepare a hydrochloric acid solution with a concentration of 5.00×10^{-4} mol dm^{-3} by diluting the sample with water.
Which of these is the correct volume of water that should be added? [1]
A 45.0 cm^3
B 95.0 cm^3
C 100 cm^3
D 995 cm^3

Answers and quick quiz 2 online

ONLINE

Summary

You should now have an understanding of:

- what is meant by the term 'mole'
- the Avogadro constant
- empirical formulae and molecular formulae
- the ideal gas equation
- how to carry out solid, gas and solution calculations
- percentage atom economy and its implications
- percentage yield
- how to write chemical equations and ionic equations

3 Bonding

Nature of ionic, covalent and metallic bonds

All types of chemical bond — whether ionic, covalent or metallic — are due to electrostatic attractions between positively charged and negatively charged particles.

Ionic bonding

REVISED

Dot-and-cross diagrams (Figure 3.1) show how atoms form ions when electrons are **transferred** from a metal atom (which becomes a positively charged ion) to a non-metal atom (which becomes a negatively charged ion).

Ionic bonding happens between oppositely charged **ions** in a lattice.

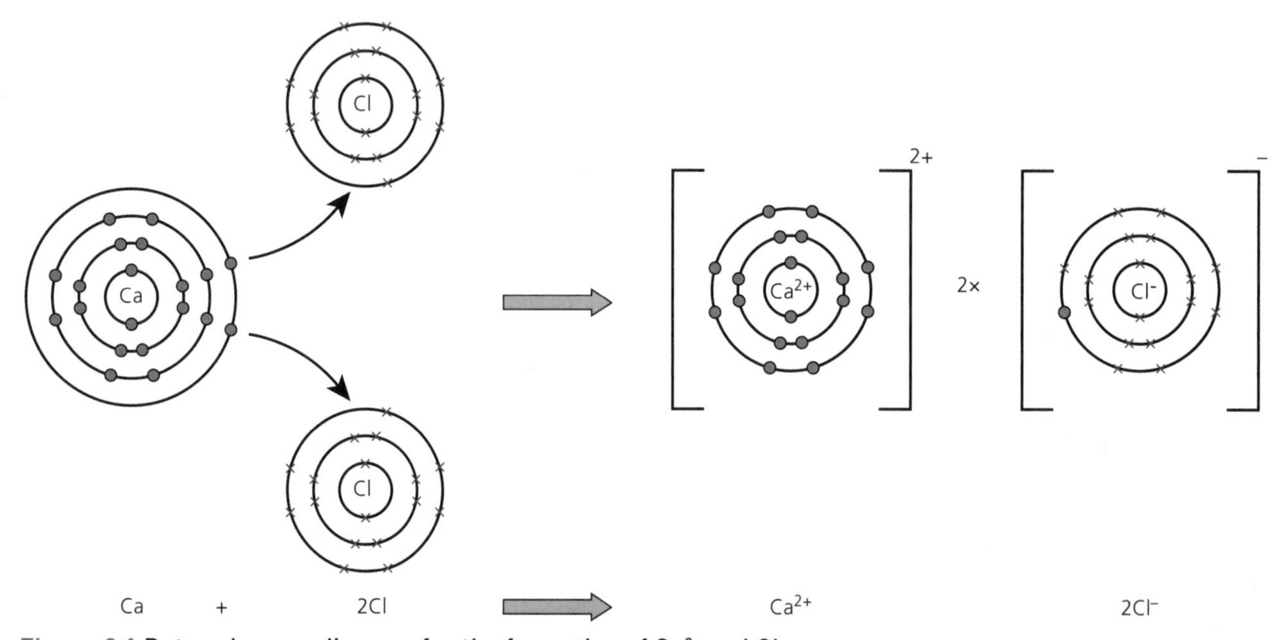

Figure 3.1 **Dot-and-cross diagram for the formation of Ca^{2+} and Cl^-**

Oppositely charged ions are attracted to each other by strong electrostatic forces. This makes it difficult to separate the ions from each other. Therefore, ionic substances:

- have high melting points
- are good electrical conductors when molten and in solution (when the ions are free to move) but poor electrical conductors when solid (when the ions are fixed in their lattice and are not free to move)

When ions pack together, they form a giant ionic structure in which positive and negatively charged ions are arranged in a lattice (Figure 3.2).

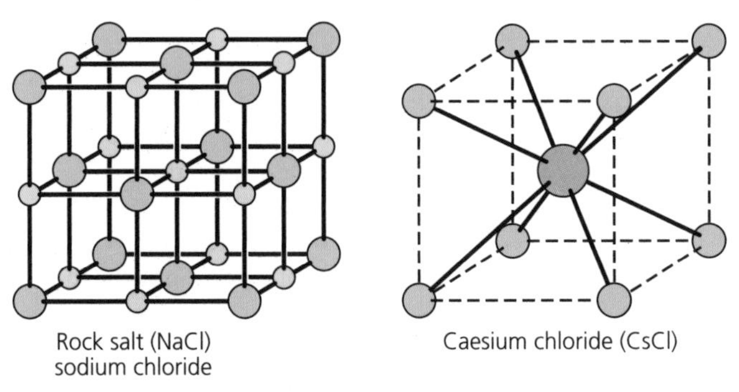

Figure 3.2 Diagrams showing part-structures for NaCl and CsCl

Typical mistake

What is wrong with this student answer? 'Sodium chloride conducts electricity when dissolved in water because electrons are free to move.'

The student should have said 'ions', not 'electrons'. This is a common error.

Also, never mention *intermolecular* forces when explaining why sodium chloride has a high melting point — mention strong electrostatic forces acting between ions instead, because there are no molecules in sodium chloride.

Now test yourself

TESTED

1 Draw dot-and-cross diagrams to show the bonding in (a) magnesium oxide and (b) sodium oxide. [Atomic number data: Mg = 12; O = 8; Na = 11]
2 Explain why lithium fluoride has a high melting point.

Answers on p. 216

Covalent bonding

REVISED

Atoms may covalently bond together to form **molecules**. These molecules may be very easy to separate because they are attracted to each other by weak intermolecular forces — these give the substance relatively low melting and boiling points. These substances have **simple covalent** structures (Figure 3.3).

A **covalent bond** involves electron pairs being *shared* between atoms.

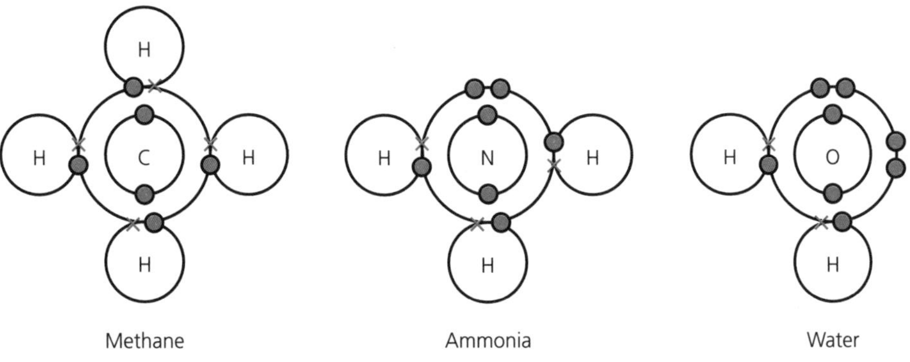

Figure 3.3 Examples of simple covalent structures

Giant molecules, like diamond (Figure 3.4), can form in which atoms are bonded throughout the structure using strong covalent bonds. This gives substances a very high melting point because these bonds will require a lot of energy to break. These are called **giant covalent (or macromolecular) structures**.

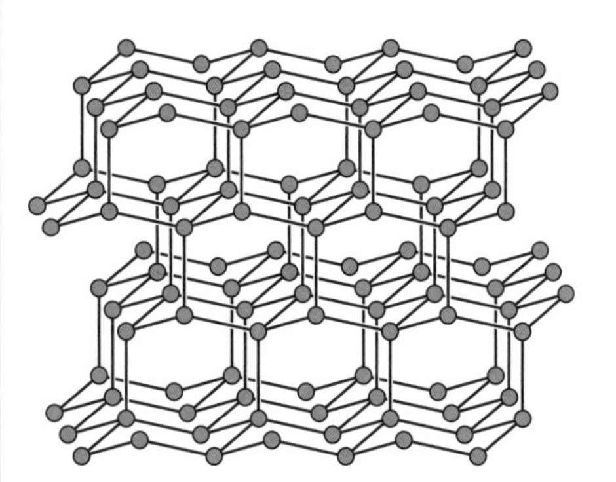

Figure 3.4 Structure of diamond

Exam tip

Normally it is true that when a metal forms a compound, ionic bonding results; and covalent bonds result when non-metals form a compound. Don't get them round the wrong way!

Now test yourself

TESTED

3 Draw dot-and-cross diagrams to show the bonding in (a) methane and (b) nitrogen.
[Atomic number data: C = 6; H = 1; N = 7]
4 Explain why methane has a low melting point.

Answers on p. 216

Revision activity

Use the internet to look up other small molecules and see if you can draw their electronic structures, showing their covalent bonds.

Dative covalent (coordinate) bonding

REVISED

In the formation of the ammonium ion, NH_4^+, the lone pair of electrons on the nitrogen atom is used to form a single covalent bond with a proton (Figure 3.5).

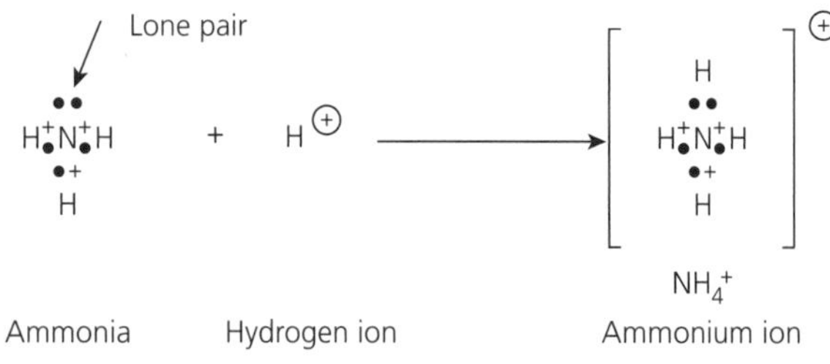

Figure 3.5 Formation of an ammonium ion, NH_4^+

A **dative covalent** or **coordinate bond** is one in which both electrons being shared in the covalent bond come from the same atom.

Carbon monoxide and ozone, O_3, are examples of molecules that contain a dative bond in which two electrons are donated from the oxygen atom.

Now test yourself

TESTED

5 Draw a dot-and-cross diagram for a carbon monoxide molecule.
[Atomic number data: C = 6; O = 8]

Answer on p. 216

Metallic bonding

REVISED

Metals consist of a lattice of **positive ions** surrounded by **delocalised electrons** (Figure 3.6).

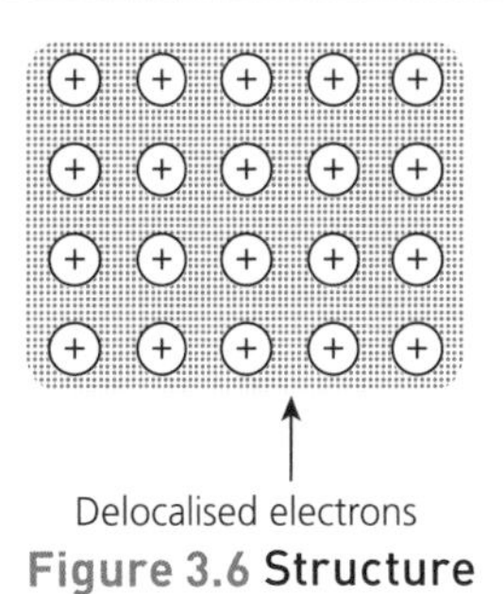

Figure 3.6 Structure of a metal

The mutual attraction of the positive ions for the delocalised electrons is a strong electrostatic force and gives most metals **high melting points** because it is difficult to separate the ions from each other.

The mobile electrons can carry an electrical charge and this explains why metals can conduct electricity so well.

The layers of metal ions within the lattice can also slide across each other fairly easily when a force is applied; this means that metals can be beaten into sheets and are said to **malleable**.

Typical mistake

What is wrong with this student's answer in explaining why the electrical conductivity of a metal is good? 'Because there are mobile ions that can carry the charge.'

They should have said that mobile *electrons* are free to carry the charge.

Bond polarity

Electronegativity

REVISED

Figure 3.7 shows the graph of electronegativity plotted against atomic number for the first 20 elements.

Electronegativity is the power of an atom to withdraw electron density from a covalent bond.

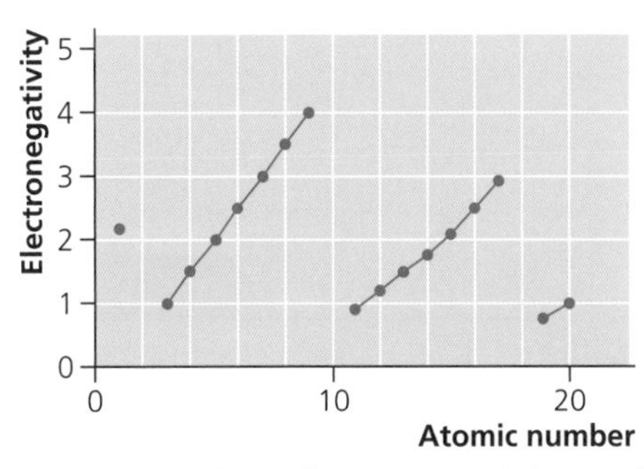

Figure 3.7 How electronegativity varies with atomic number

Electronegativity increases across a period, from left to right, and decreases going down a group. Elements with very high electronegativity values include F, N and O — these elements are very good at attracting bonded electrons towards their nuclei. This attraction causes an **asymmetrical electron distribution** in which the electronegative atom develops a slight negative charge and the other atom develops a slight positive charge. A **polar bond** is produced.

Figure 3.8 shows the electron distribution in a chlorine molecule (a non-polar molecule because both atoms are the same) and hydrogen chloride (a polar molecule due to the different electronegativities of hydrogen (2.1) and chlorine (3.0)). Hydrogen chloride is a polar molecule and has slightly charged ends: $H^{\delta+}-Cl^{\delta-}$.

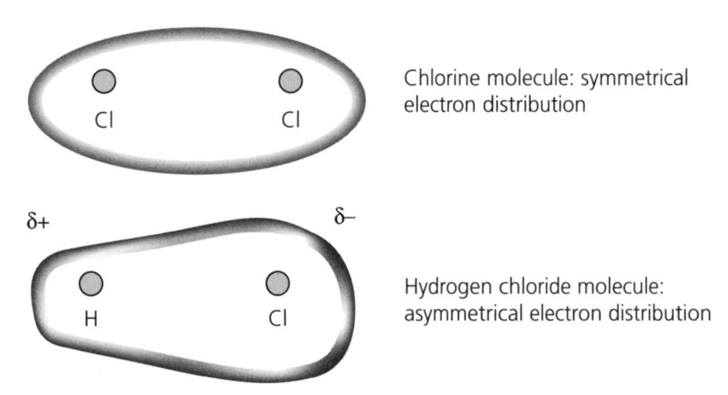

Figure 3.8 Electron distribution in chlorine and hydrogen chloride

Exam tip

Elements have different electronegativity values — this means that bonds can differ in their polarity depending on the atoms at the ends of the bond.

Now test yourself

TESTED

6 Explain why hydrogen fluoride, HF, is a polar molecule whereas hydrogen, H_2, is a non-polar molecule.
The electronegativity values of fluorine and hydrogen are 4.0 and 2.1 respectively.

Answer on p. 216

Some molecules may have polar bonds but the molecule still has no overall **dipole moment**. This is because the individual dipoles cancel out because of the three-dimensional shape of the molecule. An example of this is carbon dioxide (Figure 3.9) in which the carbon–oxygen dipoles cancel due to there being equal and opposite dipoles.

$$\overset{\delta-}{O}=\overset{\delta+}{C}=\overset{\delta-}{O}$$

Figure 3.9 Carbon dioxide molecule

Forces acting between molecules

Forces acting between molecules are called **intermolecular forces**. There are three main types and their strength decreases in the order:

hydrogen bonding > dipole–dipole forces > van der Waals forces

Molecules with permanent dipoles

REVISED

When a molecule has a permanent dipole, either dipole–dipole attractions or hydrogen bonding are possible intermolecular forces.

Hydrogen bonding

If nitrogen, oxygen or fluorine atoms are covalently bonded to hydrogen atoms in a molecule, for example in ammonia and water, an intermolecular attraction called a **hydrogen bond** can act between molecules.

A lone pair of electrons on an oxygen atom of one water molecule is attracted to the slight positive charge on a hydrogen atom in a neighbouring water molecule (Figure 3.10).

Exam tip

Forces that exist *within* a molecule are strong covalent bonds. Those *between* molecules are weak intermolecular forces.

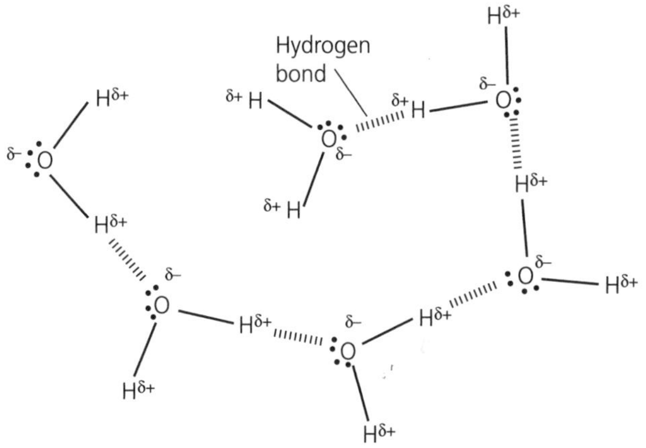

Figure 3.10 Hydrogen bonding in water

Water is a substance that has some anomalous physical properties that can be explained using hydrogen bonding:

- higher melting and boiling points than expected
- lower density as a ice than as a liquid
- the existence of surface tension

Now test yourself

TESTED

7 Draw a diagram to show the hydrogen bonding that takes place between two molecules of ammonia. Make sure that you include all lone pairs and partial charges in your diagram.

Answer on p. 216

Dipole–dipole attractions

Polar molecules that do not have N, O or F atoms bonded directly to hydrogen atoms can attract each other using **dipole–dipole attractions**. This is a weaker intermolecular force than hydrogen bonding.

Molecules of hydrogen chloride (Figure 3.11) can attract each other in this way — the slight negative charge on the chlorine atoms attracts the slight positive charge of hydrogen atoms on another molecule.

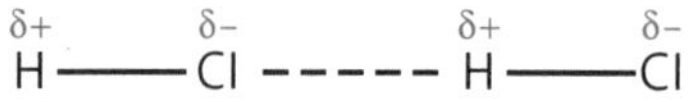

$\overset{\delta+}{H}$——$\overset{\delta-}{Cl}$ ----- $\overset{\delta+}{H}$——$\overset{\delta-}{Cl}$

Figure 3.11 Why molecules of hydrogen chloride interact

Molecules with temporary dipoles

REVISED

Many molecules are non-polar and yet they are still able to attract each other using **van der Waals forces**, the weakest of all intermolecular forces.

The electrons in all molecules are in constant motion and this movement causes 'wobbles' in electron clouds that result in temporary dipoles. This dipole may then induce another temporary dipole in a neighbouring molecule. The attraction between these temporary dipoles is called a van der Waals force.

The size of van der Waals forces depends on the number of electrons in a molecule and also the area of contact of one molecule with another.

Exam tip

The melting and boiling points of a homologous series of hydrocarbons and the elements in Group 7 (the halogens) and Group 0 (the noble gases) all increase as relative molecular mass increases. This is due to the increasing number of electrons in molecules.

Example

Explain why the boiling points of the hydrocarbons — methane, ethane, propane, butane and pentane — increase in the order written.

Answer

As the relative molecular mass increases, the number of electrons in each molecule increases. This means that more and stronger van der Waals forces exist between the molecules, and they will be more difficult to separate from each other, giving them higher boiling points.

Now test yourself TESTED

8 Explain why the boiling points of the halogens increase on descending the group.

Answers on p. 216

States of matter

Structure of substances

REVISED

Substances have different physical properties that depend on their structures (Figure 3.12).

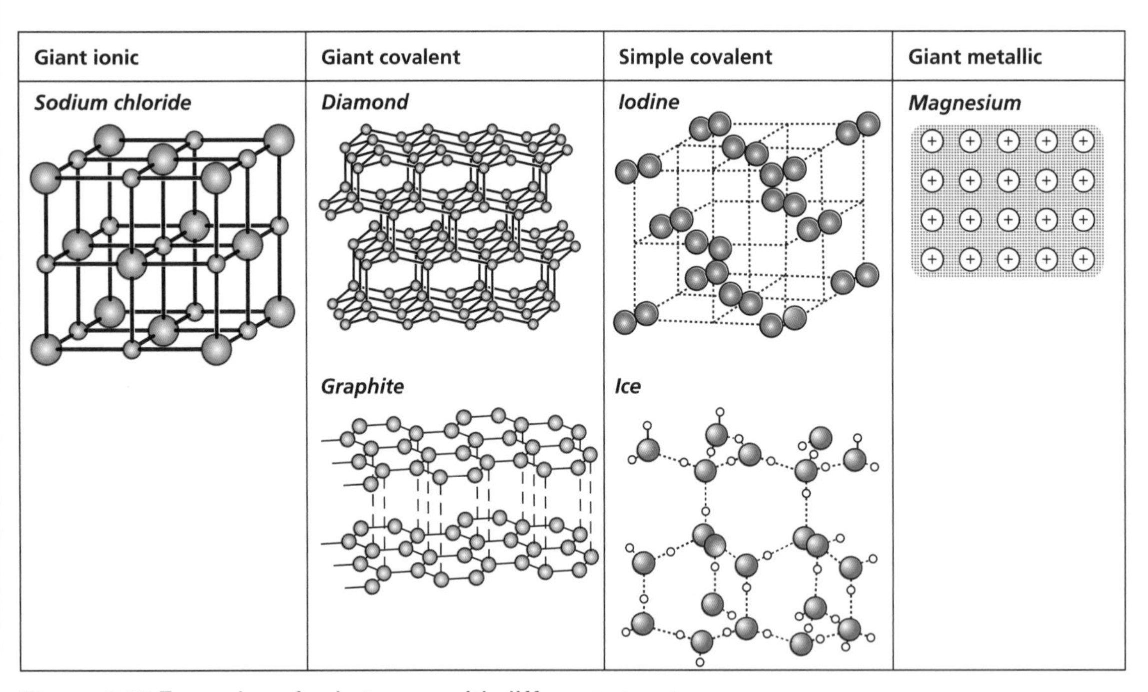

Figure 3.12 Examples of substances with different structures

Giant ionic substances have high melting and boiling points because the oppositely charged ions are attracted to each other by strong electrostatic forces. They are good electrical conductors when molten and in solution because the ions are free to carry the electrical charge. In the solid state, the ions are fixed in place in the lattice.

Giant covalent (macromolecular) substances have high melting and boiling points because the atoms are bonded throughout by strong covalent bonds. This means that a lot of energy is required to break the bonds and to melt the substance.

Simple covalent substances have low melting and boiling points because it is relatively easy to overcome the weak intermolecular forces to separate one molecule from others.

Metallic substances generally have high melting and boiling points because there are strong electrostatic forces acting between the mobile electrons and the positive ions in the lattice. They are good electrical conductors because the electrons are free to move and carry the charge.

Now test yourself

TESTED

9 Explain why sodium fluoride, NaF, has a high melting point whereas the element fluorine, F_2, has a low melting point.

Answer on p. 216

Shapes of molecules and ions

Deducing the shape of a molecule

REVISED

The three-dimensional shape of a molecule depends on the repulsion of the electron pairs around the central atom in the molecule. The decreasing order of strengths of these is as follows:

lone pair–lone pair > lone pair–bonding pair > bonding pair–bonding pair

Table 3.1 The shapes of molecules

Number of bonding pairs around the central atom (or atom–atom links)	Number of non-bonding pairs	Name of shape with diagram and examples
2	0	Linear  180° CO_2, HCN, $BeCl_2$, $[Ag(NH_3)_2]^+$
2	2	Bent 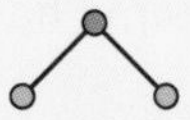 H_2O, H_2S, NO_2^-, ClO_2
3	0	Trigonal planar 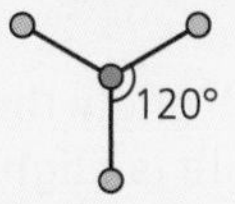 120° BF_3, NO_3^-, CO_3^{2-}, SO_3
3	1	Pyramidal 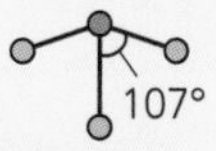 107° NH_3, PH_3, SO_3^{2-}, ClO_3^-
3	2	T-shape 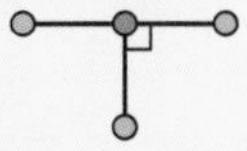 ClF_3

Number of bonding pairs around the central atom (or atom–atom links)	Number of non-bonding pairs	Name of shape with diagram and examples
4	0	Tetrahedral 109°28′ NH_4^+, CH_4, SO_4^{2-}, $CuCl_4^{2-}$
4	2	Square planar XeF_4, ICl_4^-
5	0	Trigonal bipyramidal 120° PCl_5
6	0	Octahedral SF_6, $[M(H_2O)_6]^{2+}$ where M = a metal ion

When working out the shape of a molecule or ion, you must work out the number of bonding electrons pairs and lone pairs.

Example

Deduce the shape of a hydrogen sulfide, H_2S, molecule.

Answer

- Sulfur is in Group 6 of the periodic table and so its atoms have six outer electrons.
- Each hydrogen atom shares one electron to make a single covalent bond.
- This means there are four pairs of outer electrons — two bonding pairs and two lone pairs.
- Lone pairs repel the bonding pairs slightly more strongly than the bonding pairs repel each other, so the internal angle is slightly less than the tetrahedral angle (109.5°).
- The shape name is bent (non-linear) and the internal angle is about 104.5° (Figure 3.13).

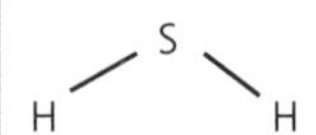

Figure 3.13 **A hydrogen sulfide molecule**

Exam tip

As a 'rule of thumb', each lone pair causes the expected tetrahedral angle to decrease by about 2.5°. So, two lone pairs of electrons would cause a decrease of about 5° from 109.5° — hence 104.5° (as in water and hydrogen sulfide).

Example

Deduce the shape of a PF_6^- ion.

Answer

- Phosphorus is in Group 5 of the periodic table and so its atoms have five outer electrons.
- Each fluorine atom shares one electron to make a single covalent bond, contributing six electrons to the outer layer.
- The negative charge on the ion adds one more electron.
- The total number of electrons around the phosphorus atom is 5 + 6 + 1 = 12, meaning six bonding pairs. There are no lone pairs.
- The shape (Figure 3.14) will therefore be octahedral and the internal bond angles will be 90° and 180°.

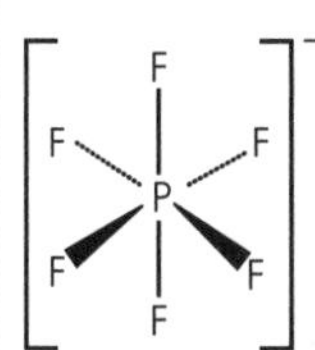

Figure 3.14 **A phosphorus(v) hexafluoride ion**

Revision activity

Draw the common shapes in Table 3.1 and use wedged and dashed bonds to indicate where the molecules are three-dimensional.

Now test yourself

TESTED

10 Sketch the shapes and indicate the internal bond angles and shape names of each of these covalent particles:
(a) NH_4^+
(b) BF_3
(c) SF_6

Answer on p. 216

Exam practice

1 (a) Describe, with a diagram, the lattice structure of sodium chloride. [2]
(b) Using your diagram from part (a), explain why sodium chloride has these properties:
(i) Sodium chloride is a good electrical conductor when molten but not when solid.
(ii) Sodium chloride has a high melting point. [2]

2 Explain the meaning of these terms:
(a) covalent bond [2]
(b) dative bond [2]
(c) giant covalent (macromolecular) structure. [2]

3 (a) Draw a dot-and-cross diagram for boron trifluoride, BF_3. [2]
(b) Comment on the outer electronic structure of the boron atom in boron trifluoride. [1]
(c) Considering your previous answers, indicate how boron trifluoride may react with ammonia. [2]

4 (a) State the strongest type of intermolecular force present in water and the strongest type of intermolecular force in hydrogen sulfide (H_2S). [2]
(b) Draw a diagram to show how two molecules of water are attracted to each other by the type of intermolecular force you stated in part (a). [3]
(c) Explain why the boiling point of water is much higher than the boiling point of hydrogen sulfide. [2]
(d) Explain why the boiling points increase from H_2S to H_2Te in Group 6. [2]

5 Which of these species has a trigonal planar structure? [1]

A PH_3
B BCl_3
C H_3O^+
D CH_3^-

6 Use your understanding of intermolecular forces to predict which of these compounds has the highest boiling point. [1]

A HF
B HCl
C HBr
D HI

7 Which type of bond is formed between N and B when a molecule of NH_3 reacts with a molecule of BF_3? [1]

A ionic
B covalent
C coordinate (dative)
D van der Waals

Answers and quick quiz 3 online

ONLINE

Summary

You should now have an understanding of:

- ionic bonding
- covalent bonding
- dative (coordinate) bonding
- metallic bonding
- bond polarity and electronegativity
- hydrogen bonding
- dipole–dipole forces
- van der Waals forces
- the relationship between physical properties and structure
- shapes of molecules and ions

4 Energetics

Enthalpy change, ΔH

An enthalpy change, ΔH, is defined as the heat energy change at constant pressure. If the enthalpy change is a standard value, with the symbol $\Delta H^{\ominus}_{298}$, then particular conditions are specified: 100 kPa pressure and a temperature of 298 K.

The sign of ΔH

REVISED

In a chemical process, heat energy can move either from the surroundings to the chemical system or vice versa. If heat energy moves from the system to the surroundings, the chemicals lose heat energy — an **exothermic reaction** results and the sign of ΔH is negative. If heat energy is transferred from the surroundings to the system, an **endothermic reaction** results and the sign of ΔH is positive.

Calorimetry

Calculations

REVISED

Calorimetry is an experimental method for determining enthalpy changes. Simple reactions can be studied in this way. When calculating the enthalpy change for a reaction, it is usually necessary to convert from a temperature change measurement recorded in an experiment (in which water is heated or cooled) into a heat energy measurement, in J. This is done using the relationship:

$$q = mc\Delta T$$

- q is the quantity of heat energy, in J
- m is the mass of the water, in g
- c is the specific heat capacity of water, 4.18 J K^{-1} g^{-1}
- ΔT is the temperature change in °C or K

Typical mistake

Many students forget that m is the mass of water and not the mass of any solid used in the experiment.

Example 1

Calculate the enthalpy change (in kJ mol^{-1}) for the following reaction of magnesium (A_r = 24.3).

2.00 g of magnesium powder is added to an excess of 25.0 cm^3 of dilute sulfuric acid solution and the temperature is found to increase by 52.4°C.

Answer

Using $q = mc\Delta T$:

m = 25.0 g; c = 4.18 J K^{-1} g^{-1}; ΔT = 52.4°C

q = 25.0 × 4.18 × 52.4 = 5476 J = 5.476 kJ

$$\text{amount of magnesium used} = \frac{2.00}{24.3} = 0.0823 \text{ mol}$$

$$\text{heat energy released} = \frac{q \text{ (in kJ)}}{\text{amount (in mol)}} = \frac{5.476}{0.0823} = 66.5 \text{ kJ mol}^{-1}$$

The reaction is exothermic, so ΔH is negative, so the enthalpy change is −66.5 kJ mol^{-1}.

Example 2

The following experiment was carried out in which a sample of ethanol (M_r = 46.0) was put into a spirit burner (Figure 4.1). The ethanol was ignited and the temperature of the 100 cm^3 of water in the metal calorimeter was found to increase from 20.1°C to 52.6°C. The mass of the spirit burner decreased from 35.64 g to 35.14 g. Calculate the enthalpy of combustion of ethanol.

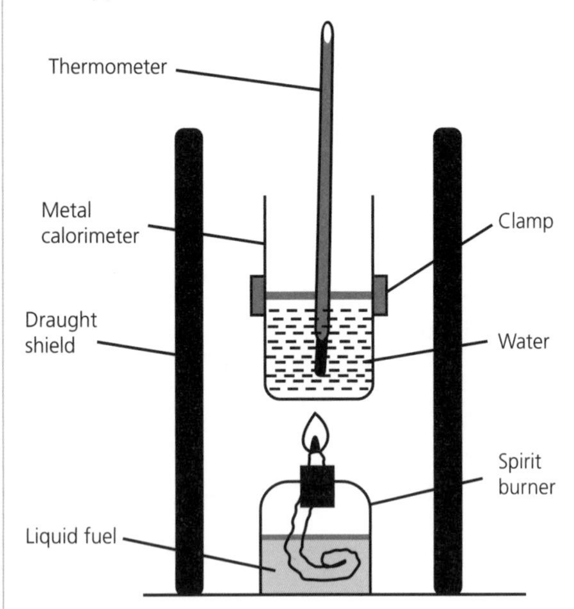

Figure 4.1 **Measuring an enthalpy of combustion**

Answer

Using $q = mc\Delta T$:

$m = 100.0\,g$; $c = 4.18\,J\,K^{-1}\,g^{-1}$; $\Delta T = 52.6 - 20.1 = 32.5°C$

$q = 100 \times 4.18 \times 32.5 = 13\,585\,J = 13.59\,kJ$

amount of ethanol used = mass of ethanol burned/M_r for ethanol

$$= \frac{35.64 - 35.14}{46.0} = 0.0109\ \text{mol}$$

$$\text{heat energy released} = \frac{q\ (\text{in kJ})}{\text{amount (in mol)}} = \frac{13.59}{0.0109} = 1247\ \text{kJ mol}^{-1}$$

The reaction is exothermic, so ΔH is negative, and the enthalpy change is $-1250\,kJ\,mol^{-1}$.

Now test yourself

TESTED

1. In an experiment, 0.95 g of powdered zinc (A_r = 65.4) is added to an excess of 25.0 cm^3 of 2.00 $mol\,dm^{-3}$ hydrochloric acid. The temperature rises by 11.4°C. Calculate the enthalpy change, in $kJ\,mol^{-1}$, for this reaction. The specific heat capacity of water is $4.18\,J\,°C^{-1}\,g^{-1}$.
2. In an experiment, 0.16 g of methanol (CH_3OH, M_r = 32.0) is burned when a spirit burner heats 100 cm^3 of water from 17.0°C to 25.0°C. Calculate the enthalpy change of combustion of methanol.

Answers on p. 217

Simple applications of Hess's law

Enthalpy of formation and combustion

REVISED

The **standard enthalpy of formation**, $\Delta_f H^\ominus$, is defined as the heat (or enthalpy) change when 1 mol of a compound is formed from its constituent elements, all reactants and products in their standard states at 298 K and 100 kPa pressure.

The standard enthalpy of formation, $\Delta_f H^\ominus$, for any *element* in its standard state is, by definition, taken as zero.

Equations to represent the enthalpy of formation of sodium chloride and of ethane gas in which 1 mol of the compound is formed are as follows:

$Na(s) + ½Cl_2(g) \rightarrow NaCl(s)$

$2C(s) + 3H_2(g) \rightarrow C_2H_6(g)$

The **standard enthalpy of combustion**, $\Delta_c H^\ominus$, is defined as the enthalpy change when 1 mol of a substance is completely burned in oxygen under standard conditions — 298 K and 100 kPa pressure — with all reactants and products in their standard states.

For example, the equation that represents the enthalpy of combustion of liquid methanol ($CH_3OH(l)$) is:

$CH_3OH(l) + 1½O_2(g) \rightarrow CO_2(g) + 2H_2O(l)$

Hess's law states that the enthalpy change of a process is independent of the route taken, whether it be direct or indirect.

Determining unknown enthalpies

REVISED

Cycle 1: using enthalpies of formation

Consider the following reaction, for which the enthalpy change is unknown:

$4NH_3(g) + 5O_2(g) \rightarrow 4NO(g) + 6H_2O(l)$

$\Delta_f H^\ominus$ values (in $kJ\,mol^{-1}$) are: $NO(g) = +90.4$; $H_2O(l) = -286$; $NH_3(g) = -46.2$.

Applying Hess's law, the enthalpy change, $\Delta H^\ominus$, for this reaction can be determined by drawing an energy cycle using the $\Delta_f H^\ominus$ values, as in Figure 4.2.

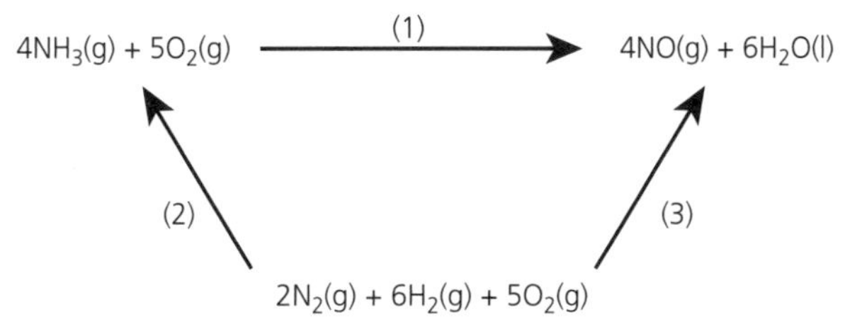

Figure 4.2

Using the cycle, it can be seen that enthalpy change (1) = −(2) + (3) = (3) − (2).

Letting $\Delta H^\ominus$ represent (1):

$$\Delta H^\ominus = 4\Delta_f H^\ominus[NO(g)] + 6\Delta_f H^\ominus[H_2O(l)] - 4\Delta_f H^\ominus[NH_3(g)] - 5\Delta_f H^\ominus[O_2(g)]$$

Substituting the appropriate given enthalpy of formation data gives:

$\Delta H^\ominus = (4 \times 90.4) + (6 \times -286) - (4 \times -46.2) - (5 \times 0) = -1170\,kJ$

Exam tip

It is sometimes more convenient to remember that if a question asks for $\Delta H^\ominus$ to be determined and only $\Delta_f H^\ominus$ data are provided, then:

$\Delta H^\ominus = \Sigma\Delta_f H^\ominus(\text{products}) - \Sigma\Delta_f H^\ominus(\text{reactants})$

Now test yourself

TESTED

3 Write balanced equations that represent the standard enthalpies of formation of these substances:
(a) carbon dioxide, $CO_2(g)$
(b) hexane, $C_6H_{14}(l)$

4 Calculate the enthalpy change for the reaction:

$P_4O_{10}(s) + 6H_2O(l) \rightarrow 4H_3PO_4(l)$

given the following standard enthalpies of formation (in $kJ\,mol^{-1}$): $P_4O_{10}(s) = -2984.0$; $H_2O(l) = -285.8$; $H_3PO_4(l) = -1279$.

Answers on p. 217

Cycle 2: Using enthalpies of combustion

Example

Determine $\Delta_f H^\ominus$ for methane, given that the enthalpies of combustion of methane, carbon and hydrogen are −890.4, −393.5 and −285.8 $kJ\,mol^{-1}$ respectively.

Answer

The equation representing the enthalpy of formation of methane is:

$C(s) + 2H_2(g) \rightarrow CH_4(g)$

Figure 4.3 shows a cycle that has this equation across the top, and the combustions down the sides.

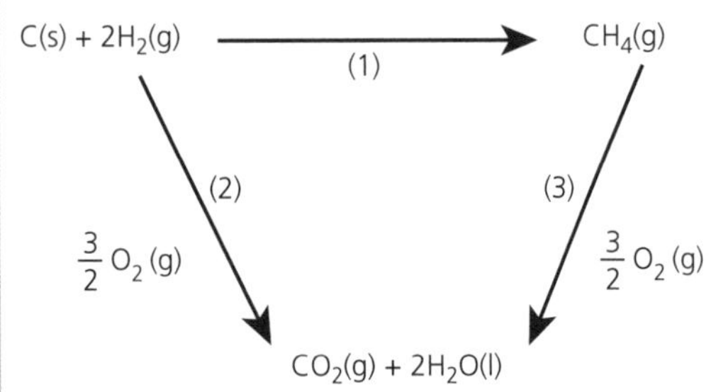

Figure 4.3

enthalpy change for the direct route from elements to oxides = (2)

enthalpy change for the indirect route from elements to oxides = (1) + (3)

Hence:

(2) = (1) + (3) or (1) = (2) − (3)

$\Delta_f H^\ominus[CH_4(g)] = \Delta_c H^\ominus[C(s)] + 2\Delta_c H^\ominus[H_2(g)] - \Delta_c H^\ominus[CH_4(g)]$

$= -393.5 + (2 \times -285.8) - (-890.4) = -74.7\,kJ\,mol^{-1}$

Now test yourself

TESTED

5 Use the data below to calculate the enthalpy of formation of propane.

$C(s) + O_2(g) \rightarrow CO_2(g) \quad \Delta H = -393.5\,kJ\,mol^{-1}$

$H_2(g) + \frac{1}{2}O_2(g) \rightarrow H_2O(l) \quad \Delta H = -285.8\,kJ\,mol^{-1}$

$C_3H_8(g) + 5O_2(g) \rightarrow 3CO_2(g) + 4H_2O(l) \quad \Delta H = -2220.0\,kJ\,mol^{-1}$

Answer on p. 217

Bond enthalpies

Mean bond energy

REVISED

A **mean bond energy** is defined as the average energy required to break 1 mol of specified bonds in the gas phase measured under standard conditions (298 K and 100 kPa pressure) over several different compounds containing the bond of interest.

- The average bond enthalpy of the N–H bond is +388 kJ mol^{-1}. This means that 388 kJ of heat energy are required to break 1 mol of N–H bonds in the gas phase. The decomposition of 1 mol of ammonia to form its constituent gaseous atoms:

  $NH_3(g) \rightarrow N(g) + 3H(g)$

  would require 3 × 388 kJ of heat energy, that is, 1164 kJ, because three N–H bonds are being broken.
- For a hydrogen molecule, the change H–H(g) → 2H(g) represents the bond energy of the H–H bond; this has a value of +436 kJ mol^{-1}. However, unlike the H–H bond, many other covalent bonds (for example C–H, C–C, C=O and N–H) vary in their strength depending on the molecule in which the bond is present. For this reason, *mean* bond energies are quoted for covalent bonds.
- Bond energies are always endothermic because energy must be provided to overcome the electrostatic attraction between the shared electrons in the covalent bond and the positively charged nuclei.

Exam tip

Experimental enthalpy values may differ from calculated mean bond energies. This is because mean bond energies are averages and not compound-specific values.

Example

Calculate $\Delta H^\ominus$ for the following reaction using mean bond energies:

$$CH_4(g) + 2Br_2(g) \rightarrow CH_2Br_2(g) + 2HBr(g)$$

Answer

The mean bond energies, in kJ mol^{-1}, are: C–H = +412; Br–Br = +193; C–Br = +276; H–Br = +366.

The energy cycle shown in Figure 4.4 can be drawn to represent the reaction.

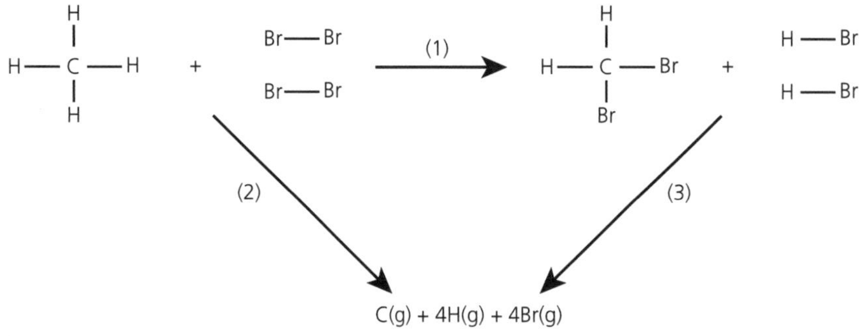

Figure 4.4

From the cycle, it can be seen that (1) = (2) – (3).

Using E to represent mean bond energy:

$$\Delta H(1) = (4 \times E[\text{C–H}]) + (2 \times E[\text{Br–Br}]) - (2 \times E[\text{C–H}]) - (2 \times E[\text{C–Br}]) - (2 \times E[\text{H–Br}])$$

Substituting the relevant bond energies gives:

$$(4 \times 412) + (2 \times 193) - (2 \times 412) - (2 \times 276) - (2 \times 366) = -74\,\text{kJ mol}^{-1}$$

Exam tip

In bond energy calculations, it is sometimes easier to remember:

$$\Delta H = \text{bond-breaking energy} - \text{bond-making energy}$$

Or simply 'break minus make'.

Now test yourself

TESTED

6 The mean bond energies N=N = 409, H–H = 436, N–H = 388 and N–N = 163 are in kJ mol^{-1}. Calculate the approximate enthalpy change, ΔH, for this reaction of hydrogen:

$$N_2H_2(g) + H_2(g) \rightarrow N_2H_4(g)$$

```
N = N        H               H     H
/     \  +   |    ------>     \   /
H      H     H                 N — N
                              /     \
                             H       H
```

Answer on p. 217

Exam practice

1 The aim of this question is to determine the enthalpy change for the reaction between 0.56 g of calcium (A_r = 40.1) and 25.0 cm^3 of 2.0 mol dm^{-3} dilute hydrochloric acid. The initial temperature of the hydrochloric acid is 21.1°C and it reaches a final temperature of 32.0°C. The specific heat capacity of water is 4.18 J K^{-1} g^{-1}.
 (a) Calculate the number of moles of calcium used. [1]
 (b) Calculate the number of moles of acid used. [1]
 (c) Write a balanced equation for the reaction. [1]
 (d) Hence show that the acid is in excess in the reaction. [1]
 (e) Calculate a value for the amount of heat involved in this experiment in kJ. [2]
 (f) Convert your answer in part (e) into a value for the enthalpy change, ΔH, in kJ mol^{-1}. [1]

2 For the reaction:

$I_2(g) + Cl_2(g) \rightarrow 2ICl(g)$ $\Delta H = -11$ kJ mol^{-1}

If the mean bond energies are E(I–I) = +158 kJ mol^{-1} and E(Cl–Cl) = +242 kJ mol^{-1}, calculate the strength of the I–Cl bond. [3]

3 Calculate the enthalpy of combustion of but-1-ene, $C_4H_8(g)$, given that the enthalpy of formations of but-1-ene, carbon dioxide and water are +1.2, –393.5 and –285.8 kJ mol^{-1} respectively. [3]

4 Table 4.1 gives some values of standard enthalpies of formation ($\Delta_f H^\ominus$).

Table 4.1

Substance	$F_2(g)$	$CF_4(g)$	$HF(g)$
$\Delta_f H^\ominus$/kJ mol^{-1}	0	–680	–269

The enthalpy change for the following reaction is –2889 kJ mol^{-1}.

$C_2H_6(g) + 7F_2(g) \rightarrow 2CF_4(g) + 6HF(g)$

Use this value and the standard enthalpies of formation in Table 4.1 to calculate the standard enthalpy of formation of $C_2H_6(g)$. [3]

Answers and quick quiz 4 online

ONLINE

Summary

You should now have an understanding of:

- what is meant by 'enthalpy change'
- how to carry out simple experiments to enable enthalpy changes to be measured
- Hess's law
- enthalpy of formation
- enthalpy of combustion
- how to apply Hess's law by drawing simple energy cycles
- mean bond enthalpies and using them to calculate enthalpy changes

5 Kinetics

Rate of reaction is defined as the *rate of change of concentration with time*. It can be expressed mathematically in the form:

$$\text{rate of reaction (mol dm}^{-3}\text{ s}^{-1}) = \frac{\text{change in concentration (mol dm}^{-3})}{\text{time for concentration change (s)}}$$

Collision theory

Effective collisions

REVISED

For reactions to occur, particles with sufficient energy must collide — the type of collision that leads to a reaction is called an **effective collision** (Figure 5.1b). If particles do not have enough energy for a reaction, or do not collide with the correct alignment or orientation, an **ineffective** or **inelastic collision** may result (Figure 5.1a).

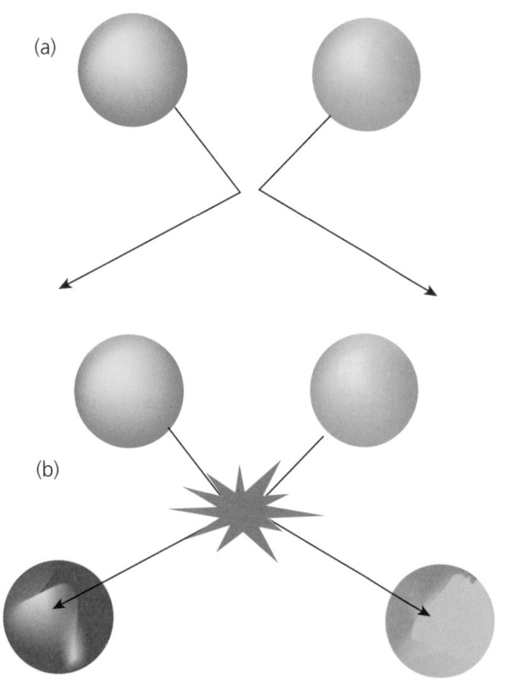

Figure 5.1 Collisions in reactions: (a) ineffective; (b) effective

The **activation energy**, E_a, is the minimum energy required for a reaction to take place. If the combined energy of the particles colliding is less than the activation energy, then a reaction is not likely to occur — this is an ineffective collision. If the combined energy of the particles exceeds the activation energy then an effective collision is more likely.

Most collisions that take place between particles do not lead to a reaction because the combined energy of the particles is lower than the activation energy.

Factors affecting rate of reaction

Changing the temperature

REVISED

As the temperature increases, so too does the average kinetic energy of the particles in a reaction mixture. Particles will collide with a greater combined energy and if this exceeds the reaction activation energy then a reaction may occur.

The distribution of particles at two different temperatures, T_1 and T_2, can be represented on a **Maxwell–Boltzmann distribution** (Figure 5.2).

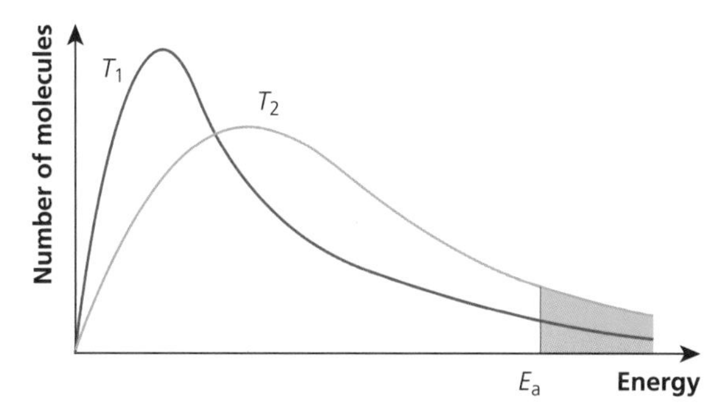

Figure 5.2 Maxwell–Boltzmann distributions at different temperatures

- At the lower temperature, T_1, the proportion of molecules with an energy greater than the activation energy, E_a, is given by the blue-shaded area.
- If the temperature is increased to T_2, a new distribution is formed in which the peak of the graph shifts to higher energy. A *much* greater proportion of particles possess an energy greater than the activation energy — the pink+blue-shaded areas — and therefore there will be a lot more successful collisions at the higher temperature, so the rate of reaction increases dramatically.
- Rate of reaction is normally very sensitive to temperature increases. This is because the proportion of molecules having an energy exceeding the activation energy — the area to the right of E_a — increases dramatically (exponentially) as the temperature increases. A small temperature increase can cause a significant shift of the Maxwell–Boltzmann distribution to the right, therefore moving more molecules to higher energies.

Typical mistake

Many students state that it is the fact that molecules collide more often at a higher temperature that causes the greater rate of reaction. It is because more molecules are colliding with greater *energy*, and therefore sufficient energy to react, that has the much greater effect.

Now test yourself

TESTED

1 The rate of reaction between hydrogen gas and oxygen gas to form water is sensitive to changes in temperature.

$2H_2(g) + O_2(g) \rightarrow 2H_2O(l)$

Suggest why an increase in temperature increases the rate of this reaction.

Answer on p. 217

Changing the concentration

REVISED

Concentration is a measure of the number of particles per unit volume and is normally expressed in $mol\,dm^{-3}$. The greater the concentration, the higher the number of particles per unit volume.

As the concentration of a solution increases, there will be more particles in the same volume. More collisions per unit time will occur and a greater number of effective collisions will take place, giving a faster rate of reaction.

In the left-hand box in Figure 5.3, the concentration of red particles is much lower than that on the right-hand side. This means that fewer collisions place per unit time between the red and blue balls on the left-hand side than on the right. This would give a lower rate of reaction because fewer effective collisions would result.

Exam tip

With gases, as the pressure is increased the concentration of particles also increases. This is analogous to increasing the concentration of a solution.

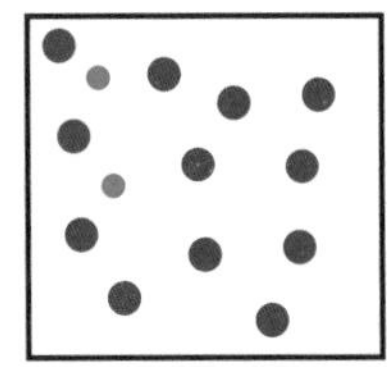

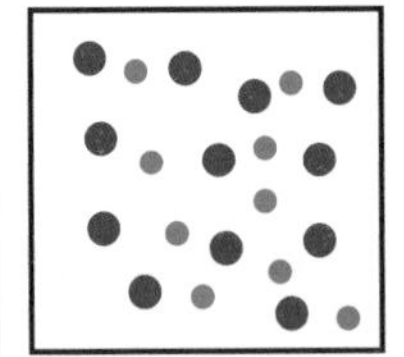

Figure 5.3 The effect of concentration

Using a catalyst

REVISED

Manganese(IV) oxide is a **catalyst** in the decomposition of hydrogen peroxide:

$$2H_2O_2(aq) \rightarrow 2H_2O(l) + O_2(g)$$

Without the catalyst, the reaction rate would be significantly slower. When manganese(IV) oxide is added, it provides an alternative reaction pathway, or mechanism, by which the reaction can take place. This alternative route is faster, because it has a lower activation energy.

If the activation energy is lowered (Figure 5.4) there will be a greater proportion of particles with an energy exceeding the old activation energy threshold and a faster reaction rate results.

A **catalyst** is a substance that increases the rate of a reaction by *providing an alternative reaction pathway* with a *lower activation energy*. At the end of the reaction a catalyst is chemically unchanged.

Exam tip

Note that the activation energy is lowered using a catalyst. However, when explaining the effect of temperature and concentration on rate, the activation energy stays the same.

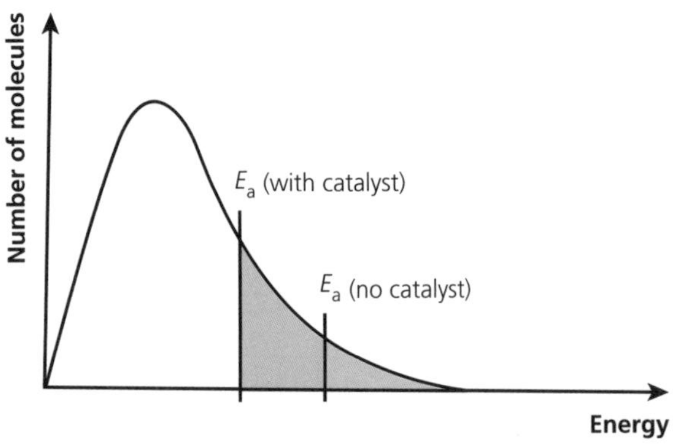

Figure 5.4 The effect of a catalyst on activation energy

You can also sketch an energy profile for a reaction to show how a catalyst affects the activation energy for a process, as shown in Figure 5.5. Note the following about this diagram:

- The activation energy for the uncatalysed reaction is much higher than that for the catalysed reaction.

- The enthalpy change for the reaction is unchanged — a catalyst affects the rate but not the amount of heat absorbed or released.

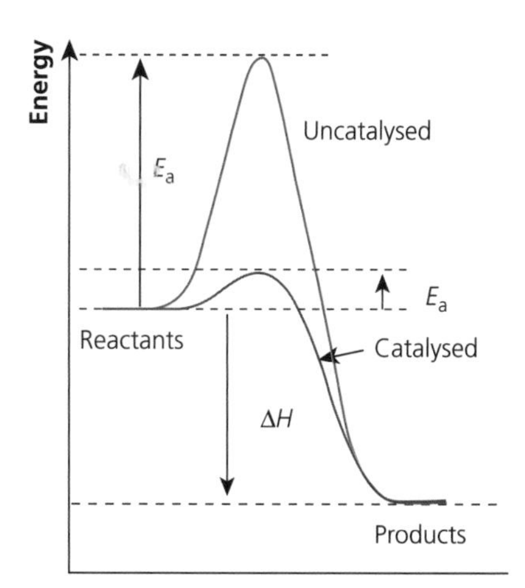

Figure 5.5 Energy profile for a catalysed reaction

Now test yourself

TESTED

2 The reaction between sodium thiosulfate solution, $Na_2S_2O_3(aq)$, and dilute hydrochloric acid can be represented by:

$Na_2S_2O_3(aq) + 2HCl(aq) \rightarrow 2NaCl(aq) + S(s) + SO_2(g) + H_2O(l)$

Explain why the rate of this reaction increases when the concentration of sodium thiosulfate increases.

3 Give an example of a catalyst in use, and explain how it increases the rate of the chemical reaction.

Answers on p. 217

Exam practice

1 (a) Give the meaning of the term 'activation energy'. [1]

Figure 5.6 shows a distribution of molecular energies at a particular temperature. The activation energy is also shown for the reaction.

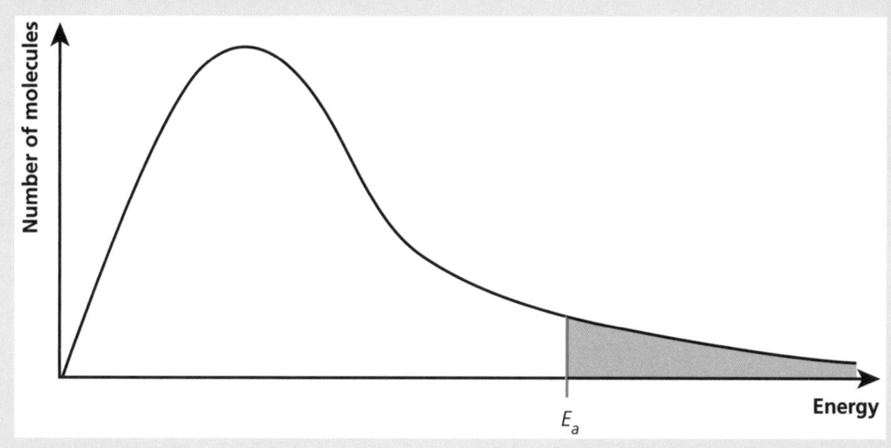

Figure 5.6

(b) Label on the graph:

- the most common energy [1]
- molecules with the highest energy [1]
- molecules with the lowest energy [1]

(c) On the same axes, draw the graph expected when the reaction mixture is heated by a further 10°C. [2]

(d) Use your graphs to explain why an increase in temperature of 10°C results in an average increase in the rate of reaction by a factor of approximately 2. [2]

2 For the reaction $2HI(g) \rightarrow H_2(g) + I_2(g)$, the activation energy when uncatalysed is 183 kJ mol^{-1} and when catalysed with gold it is 105 kJ mol^{-1}. ΔH for the reaction is −52 kJ mol^{-1}.

(a) Sketch an energy profile diagram for the reaction, including the activation energies for both the uncatalysed and catalysed reactions. [2]

(b) Calculate the activation energy for the reverse reaction, $H_2(g) + I_2(g) \rightarrow 2HI(g)$, in both the uncatalysed and catalysed reactions. [2]

(c) Explain why increasing the concentration of hydrogen iodide gas results in a faster reaction rate. [2]

Questions 3 and 4 are about the Maxwell–Boltzmann distribution shown in Figure 5.7.

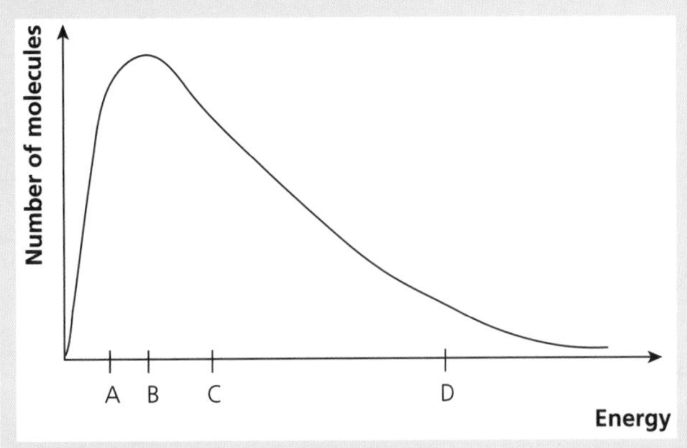

Figure 5.7

3 Which letter best represents the mean energy of the molecules? [1]

4 What does the area under the curve represent? [1]

A the total energy of the particles
B the total number of particles
C the number of particles that can react with each other
D the total number of particles that have activation energy

Answers and quick quiz 5 online

ONLINE

Summary

You should now have an understanding of:

- collision theory:
  - particles have to collide to react
  - particles need to possess a certain minimum energy to react
- activation energy
- how temperature affects the rate of reaction
- the Maxwell–Boltzmann distribution
- how concentration affects the rate of reaction
- what is meant by 'catalysts' and how they work

6 Equilibria

The dynamic nature of equilibria

Reversible reactions

REVISED

There are many reactions that can be described as being 'reversible' — that is, they can proceed in either the forward or reverse directions. For example, ethene can be hydrated to form ethanol, C_2H_5OH:

$$C_2H_4(g) + H_2O(g) \rightarrow C_2H_5OH(l)$$

On the other hand, ethanol can be dehydrated to form ethene and water:

$$C_2H_5OH(l) \rightarrow C_2H_4(g) + H_2O(g)$$

When a reaction that is reversible takes place, both the forward and reverse processes will occur until a **state of dynamic chemical equilibrium** is attained. When the system (reaction) reaches this stage, the rates of the forward and reverse reactions are equal.

When at equilibrium ($\rightleftharpoons$) the reaction above can be represented as:

$$C_2H_4(g) + H_2O(g) \rightleftharpoons C_2H_5OH(g)$$

Another example of making an alcohol in an industrial process is when carbon monoxide reacts with hydrogen to form methanol, CH_3OH:

$$CO(g) + 2H_2(g) \rightleftharpoons CH_3OH(g)$$

A state of equilibrium will occur only if the system is **closed**, so that nothing is allowed to enter or leave the reaction.

Exam tip

When a reaction is at equilibrium, the concentrations of reactants and products are not necessarily equal, but the *rates* of the forward and reverse processes are equal.

Exam tip

Both ethanol and methanol are important fuels, and both are formed industrially in equilibrium processes.

The equilibrium constant, K_c

REVISED

For a reaction at equilibrium, a mathematical expression may be written that relates the equilibrium concentrations of reactants and products to a constant called the equilibrium constant, known as $\mathbf{K_c}$.

When writing an expression for K_c:

- Focus on the product side of the equilibrium first and write each substance as a concentration, using square brackets.
- If there is more than one product, multiply the different concentrations together.
- Raise each concentration to a power, where this power is equal to the number before the substance in the chemical equation.
- Do the same with the reactants, but write this under the products as a division.

For example, in the equilibrium:

$$C_2H_4(g) + H_2O(g) \rightleftharpoons C_2H_5OH(g)$$

the equilibrium constant is given by:

$$K_c = \frac{[C_2H_5OH(g)]}{[C_2H_4(g)][H_2O(g)]}$$

For the reaction below:

$$CO(g) + 2H_2(g) \rightleftharpoons CH_3OH(g)$$

the equilibrium constant is given by:

$$K_c = \frac{[CH_3OH(g)]}{[CO(g)][H_2(g)]^2}$$

What does the equilibrium constant tell us?

The greater the value of K_c, the higher the proportion of products compared with reactants in the mixture at equilibrium, and vice versa. For example, K_c for a reaction may have a very large value of, say, 10^6. This indicates that products dominate in the equilibrium mixture and that the equilibrium position lies to the right-hand side.

Conversely, the value for K_c could be very small, for example, 10^{-6}. This means that reactants dominate the mixture, and the equilibrium position lies largely to the left-hand side.

What doesn't the equilibrium constant tell us?

The rate of reaction is unknown. An equilibrium constant gives us information about the **final composition** of the equilibrium mixture but not about how long it takes to form. It could take many years to come to equilibrium, or a fraction of a second.

Now test yourself

TESTED

1 Write down equilibrium constant expressions for the following reactions at equilibrium.
 (a) $N_2(g) + 3H_2(g) \rightleftharpoons 2NH_3(g)$
 (b) $H_2(g) + I_2(g) \rightleftharpoons 2HI(g)$
 (c) $N_2(g) + O_2(g) \rightleftharpoons 2NO(g)$

Answer on p. 217

Qualitative effects of changes in external conditions

Equilibrium position

REVISED

Reactions that come to equilibrium will consist of varying relative proportions of reactants and products. If a reaction comes to equilibrium and the quantity of reactants is greater than the quantity of products, the reaction is said to have an equilibrium position that lies on the left-hand side.

The **equilibrium position** therefore gives a measure of the extent to which the reaction takes place, and the position of equilibrium can be 'shifted' by changing the external conditions.

Le Chatelier's principle

REVISED

Using **Le Chatelier's principle**, it is possible to predict the qualitative effect of changing external conditions — for example concentration, pressure or temperature — on an equilibrium position.

Le Chatelier's principle states that if the conditions under which an equilibrium exists are changed, the position of equilibrium alters in such a way as to oppose the change in conditions.

Changing the concentration of a reactant or a product

If the concentration of a substance changes, the reaction shifts position so as to oppose the concentration change.

Consider this example. The following process involving the formation of the ester methyl methanoate (CH_3OCHO) has reached equilibrium:

$$CH_3OH(l) + HCOOH(l) \rightleftharpoons CH_3OCHO(l) + H_2O(l)$$

- If the concentration of methanol, CH_3OH, is increased by adding more moles of CH_3OH, the position of equilibrium position moves to lower the concentration of the CH_3OH — it shifts to the *right-hand side* to produce more ester.
- If methanoic acid, HCOOH, is removed from the equilibrium by reducing its concentration, the equilibrium position shifts so as to produce more HCOOH — the equilibrium shifts to the *left-hand side.*

Changing the pressure

If a reaction involves gases, the equilibrium position will be affected by changes in external pressure. Consider the manufacture of ammonia:

$$N_2(g) + 3H_2(g) \rightleftharpoons 2NH_3(g)$$

- There is a total of 4 mol of gas on the left-hand side of the equilibrium and only 2 mol of gas on the right-hand side, so the left-hand side (the reactants) will exert a greater pressure relative to the products.
- If the external pressure is increased, the equilibrium position moves so as to decrease the pressure. The reaction will do this by shifting to the right-hand side to produce more molecules of ammonia and fewer molecules of nitrogen and hydrogen.

However, if the reaction being considered is:

$$H_2(g) + I_2(g) \rightleftharpoons 2HI(g)$$

- A change in external pressure will not affect the equilibrium position because there are equal numbers of moles of gas on each side of the equation.
- The *rate* at which equilibrium is attained will be affected because pressure affects the relative distance between molecules in the gas phase. The collision rate will therefore change and the number of effective collisions will also change.

Changing the temperature

To predict how a change of temperature affects the equilibrium position, the sign of the enthalpy change, ΔH, for the reaction has to be known. Consider the reaction:

$$N_2(g) + O_2(g) \rightleftharpoons 2NO(g) \qquad \Delta H = +180.5\,kJ\,mol^{-1}$$

- The *forward* reaction is *endothermic*. Therefore, formation of NO(g) is accompanied by the removal of heat from the surroundings.

- The *reverse* process is *exothermic*. Heat will be released when N_2(g) and O_2(g) are formed.
- If the external temperature is increased, the reaction will try to remove excess heat by using its forward process and more NO(g) will form.
- Temperature also affects the rates of both the forward and the reverse process (but not equally). If the temperature is increased, the rates of both processes increase and the rate at which the equilibrium is attained also increases.

Catalysts

REVISED

A **catalyst** will not affect the equilibrium *position*; it will only affect the *rate* at which the equilibrium is attained.

The ester ethyl ethanoate, $CH_3COOC_2H_5$, is formed in this reaction:

$$C_2H_5OH(l) + CH_3COOH(l) \rightleftharpoons CH_3COOC_2H_5(l) + H_2O(l)$$

- It can take a long time for this reaction to come to equilibrium — many months, or even years.
- If a catalyst such as concentrated sulfuric acid is added, the equilibrium is attained faster.
- The rates of the forward and reverse processes are increased equally by the catalyst.

Exam tip

The equilibrium constant, K_c, gives a measure of the resulting equilibrium composition. Changes in concentration, pressure or surface area of a solid, or adding a catalyst, have no effect on the value of K_c. Only temperature will affect its value.

The importance of equilibria in industrial processes

Ethanol production

REVISED

One very important example of a reversible reaction that comes to a state of dynamic equilibrium is in the industrial manufacture of ethanol from steam and ethene:

$$C_2H_4(g) + H_2O(g) \rightleftharpoons C_2H_5OH(g) \; \Delta H \text{ -ve}$$

Ethene reacts with steam in the presence of a phosphoric acid catalyst (on a silica support) at 300°C, and a pressure of about 60–70 atm.

The process is reversible and so it is important to appreciate how conditions are changed so that a greater yield of ethanol is formed at an appreciable rate.

Pressure

- **Yield:** in the equation for the process, there are fewer gas molecules (low pressure) on the product side, and more gas molecules (high pressure) on the reactant side. When the external pressure is increased, the equilibrium will shift to the right-hand side to reduce the pressure, so a greater yield of ethanol will form.

- **Rate:** when gases are compressed, there will be more collisions per unit time because the molecules will be closer together. The rate at which equilibrium is achieved therefore increases.

Temperature

- **Yield:** decreasing the temperature will increase the yield of ethanol, because the forward exothermic reaction will be favoured.
- **Rate:** a lower temperature will result in a lower rate of reaction.

A compromise is found between yield and rate by adopting a moderate temperature.

Now test yourself

TESTED

2 Indicate the effect on (a) the equilibrium position and (b) the rate at which equilibrium is attained when the temperature is increased in this equilibrium:

$2NO_2(g) \rightleftharpoons N_2O_4(g)$

The forward reaction is exothermic.

(c) Write an expression for K_c for the equilibrium.

Answer on p. 217

Exam tip

Yield and rate are different features of an equilibrium system — Le Chatelier's principle can be used to explain the changes to the yield; collision theory explains the changes to the rate.

Catalyst

The phosphoric acid catalyst used in ethanol production affects the rate at which the equilibrium is attained; it does not affect the yield of ethanol formed.

Now test yourself

TESTED

3 Copy and complete Table 6.1 to describe some changes to the following equilibrium:

$CO(g) + 2H_2(g) \rightleftharpoons CH_3OH(g)$

The forward reaction is exothermic.

Table 6.1

	Equilibrium position	Rate at which equilibrium is attained
Total pressure is increased		
Temperature is increased		
A catalyst is added		

Answer on p. 217

Exam practice

1 In the Haber process, nitrogen and hydrogen react in the presence of an iron catalyst to form ammonia according to the equation:

$N_2(g) + 3H_2(g) \rightleftharpoons 2NH_3(g)$

(a) The production of ammonia is accompanied by heat being released. Explain whether a low temperature or a high temperature would result in the higher yield of ammonia being formed. [2]

(b) State what happens to the rate at which equilibrium is attained when:

(i) the temperature is increased [1]

(ii) the pressure is decreased [1]

(c) Explain your answers to part (b) in terms of energies of particles. [2]

(d) Explain why a moderately high temperature of 450°C is used when manufacturing ammonia industrially. [2]

(e) Explain why pressures of more than 200–350 atm are rarely used in the Haber process. [2]

(f) In one particular year, the USA produced 8 million tonnes of ammonia at a cost of $200 per tonne.

(i) Calculate the total cost of the US production of ammonia. [1]

(ii) Calculate the mass of nitrogen gas required to produce 8 million tonnes of ammonia. [A_r data: N = 14; H = 1] [2]

(g) Given that the process produces an overall ammonia yield of 15%, calculate the mass of nitrogen required when this is taken into account. [1]

2 Colourless solutions of X(aq) and Y(aq) react to form an orange solution of Z(aq) according to the following equation:

$X(aq) + 2Y(aq) \rightleftharpoons Z(aq) \qquad \Delta H = -20\,kJ\,mol^{-1}$

A student added a solution containing 0.50 mol of X(aq) to a solution containing 0.50 mol of Y(aq) and shook the mixture. After 30 seconds, there was no further change in colour. The amount of Z(aq) at equilibrium was 0.20 mol.

(a) Deduce the amounts of X(aq) and Y(aq) at equilibrium. [2]

(b) The student prepared another equilibrium mixture in which the equilibrium concentrations of X and Z were: X(aq) = 0.40 mol dm^{-3} and Z(aq) = 0.35 mol dm^{-3}. For this reaction, the equilibrium constant K_c = 2.9 mol^{-2} dm^{6}. Calculate a value for the concentration of Y at equilibrium. Give your answer to the appropriate number of significant figures. [3]

(c) The student added a few drops of Y(aq) to the equilibrium mixture of X(aq), Y(aq) and Z(aq) in part (b). Suggest how the colour of the mixture changed. Give a reason for your answer. [3]

Answers and quick quiz 6 online

ONLINE

Summary

You should now have an understanding of:

- reversible reactions
- what is meant by 'state of equilibrium'
- what is meant by 'dynamic equilibrium'
- how Le Chatelier's principle can be used to predict the direction in which an equilibrium shifts when external conditions are changed
- writing an expression for K_c for a given equilibrium
- the production of ethanol in terms of how to change external conditions to optimise both yield and rate

7 Redox reactions

Oxidation and reduction

Redox reaction

REVISED

In a **redox reaction**, one substance is **oxidised** and another is **reduced**. The substance that oxidises the other is called an **oxidising agent**. The substance that reduces the other substance is called the **reducing agent**.

A simple redox reaction takes place when a metal compound, like aluminium oxide, is electrolysed. These processes take place at the electrodes:

- at the (−) cathode: $Al^{3+} + 3e^- \rightarrow Al$
- at the (+) anode: $2O^{2-} \rightarrow O_2 + 4e^-$

Aluminium ions gain electrons, so they are **reduced**.

Oxide ions lose electrons, so they are **oxidised**.

Oxidation is the loss of electrons.

Reduction is the gain of electrons.

Oxidation states

Working out the oxidation state

REVISED

An oxidation state is a number indicating the 'formal' charge that an element would have in a compound if the compound were ionic. The oxidation state of an element is zero.

Working out the oxidation state for an element in a compound assumes that all the other oxidation states are known. The sum of all of the oxidation states in a compound is zero.

Some typical oxidation states:

- hydrogen (except when in hydrides) = +1
- Group 1 metals = +1
- Group 2 metals = +2
- oxygen (except when in hydrogen peroxide) = −2
- fluorine = −1

Example

Work out the oxidation state of the following:

(a) iron in Fe_2O_3

(b) manganese in $KMnO_4$

(c) chromium in $Na_2Cr_2O_7$

Answer

(a) In Fe_2O_3 the known oxidation state is oxygen at −2.
Therefore, 2Fe + 3(−2) = 0
2Fe = +6, so each iron = +3
This compound is called iron(III) oxide, where 'III' is the oxidation state of the iron in the compound.

(b) In $KMnO_4$ the known oxidation states are potassium at +1 and oxygen at −2.
Therefore, (+1) + Mn + 4(−2) = 0
Mn = (+8) − 1, so each manganese = +7
The compound is called potassium manganate(VII), where 'VII' is the oxidation state of manganese in the compound.

(c) In $Na_2Cr_2O_7$ the known oxidation states are sodium at +1 and oxygen at −2.
Therefore, 2(+1) + 2Cr + 7(−2) = 0
2 + 2Cr − 14 = 0, so 2Cr = 14 − 2
Cr = +6, so each chromium = +6
This compound is called sodium dichromate(VI), where 'VI' is the oxidation state of chromium in the compound.

For complex ions such as SO_4^{2-}, the sum of the oxidation states is equal to the charge on the ion.

Example

What is the oxidation state of aluminium in the ion $[AlF_6]^{3-}$?

Answer

Known oxidation states: F = −1

Therefore, Al + 6(−1) = −3

Al − 6 = −3

So Al = −3 + 6 = +3

Now test yourself

TESTED

1 What is the oxidation state of the named element in each of these compounds and ions?

(a) cobalt in $CoCl_3$

(b) chlorine in NaOCl

(c) titanium in $TiCl_4$

(d) iron in Na_2FeO_4

(e) sulfur in H_2SO_4

(f) iodine in IO_3^-

(g) manganese in MnO_4^{2-}

2 Name, using oxidation states, the compounds in question 1.

Answers on p. 217

Using oxidation states

REVISED

When carrying out a reaction, it is possible to deduce whether the reaction is redox or not, and if it is, which elements have been oxidised and which reduced.

Exam tip

If the oxidation state of an element increases, this is an **oxidation**. If it decreases this is a **reduction**.

Example

Determine whether the following are redox reactions or not by deducing the oxidation state changes.

(a) $H_2SO_4(l) + NaCl(s) \rightarrow NaHSO_4(s) + HCl(g)$
(b) $2SO_2(g) + O_2(g) + 2H_2O(l) \rightarrow 2H_2SO_4(aq)$

Answer

(a) H_2SO_4: hydrogen = +1; oxygen = −2, so sulfur = +6
NaCl: sodium = +1; chlorine = −1
$NaHSO_4$: sodium = +1; hydrogen = +1; oxygen = −2, so sulfur = +6
HCl: hydrogen = +1; chlorine = −1
So all the oxidation states before and after the reaction are the same, so this reaction is not a redox reaction.

(b) SO_2: sulfur = +4; oxygen = −2
O_2: oxygen = 0
H_2O: hydrogen = +1 and oxygen = −2
H_2SO_4: hydrogen is +1, sulfur is +6 and oxygen is −2.
Hydrogen's oxidation state stays the same.
Sulfur changes oxidation state from +4 to +6. It has been oxidised.
Oxygen (starting in O_2) changes from 0 to −2. It has been reduced.
The reaction is therefore a redox reaction.

Now test yourself

TESTED

3 Deduce the names of the oxidising and reducing agents in the reaction:

$$2SO_2(g) + O_2(g) + 2H_2O(l) \rightarrow 2H_2SO_4(aq)$$

4 Show that the reaction below is a redox reaction by determining the oxidation states of the elements that change. Also deduce the names of the oxidising and reducing agents:

$$PbO_2(s) + 4HCl(aq) \rightarrow PbCl_2(aq) + 2H_2O(l) + 2Cl_2(g)$$

Answers on p. 217

Redox equations

Writing half-equations

REVISED

Redox reactions consist of two separate processes — an oxidation and a reduction. These can be written separately in the form of **half-equations**.

Example

Magnesium metal is added to copper(II) sulfate solution to form copper and magnesium sulfate solution. Write a balanced symbol equation for the reaction and also the half-equations. Then determine which element has been oxidised and which has been reduced.

Answer

$$Mg(s) + CuSO_4(aq) \rightarrow MgSO_4(aq) + Cu(s)$$

In this reaction, the magnesium atoms lose two electrons to form magnesium ions, Mg^{2+}.

Half-equation: $Mg(s) \rightarrow Mg^{2+}(aq) + 2e^-$

The magnesium atoms have therefore been oxidised.

The copper(II) ions, Cu^{2+}, gain two electrons to form copper atoms, Cu.

Half-equation: $Cu^{2+}(aq) + 2e^- \rightarrow Cu(s)$

The copper(II) ions have therefore been reduced.

Now test yourself

TESTED

5 Write two half-equations that take place in the reactions between:
  (a) zinc and iron(II) sulfate solution
  (b) aluminium and copper(II) sulfate solution

Answers on p. 218

The two half-equations in the example above can be combined to form an overall **ionic equation**:

$$Cu^{2+}(aq) + Mg(s) \rightarrow Mg^{2+}(aq) + Cu(s)$$

When oxidation and reduction half-equations are combined, the electrons *must* cancel out.

In this reaction, magnesium has been oxidised and the copper(II) ion is the **oxidising agent**, because it caused this oxidation. Copper(II) ions are reduced, and so magnesium is called the **reducing agent**.

Example

Given the following half-equations for a redox reaction between chlorine and bromide ions, combine them together to produce an overall ionic equation:

$$Cl_2 + 2e^- \rightarrow 2Cl^-$$

$$2Br^- \rightarrow Br_2 + 2e^-$$

Answer

Adding these two together, and cancelling the electrons on each side, gives:

$$Cl_2 + 2Br^- \rightarrow Br_2 + 2Cl^-$$

Example

When lithium metal is added to water, reactions occur according to these two half-equations:

$Li \rightarrow Li^+ + e^-$

$2H_2O + 2e^- \rightarrow 2OH^- + H_2$

What is the balanced equation for the reaction?

Answer

Multiplying the first equation by 2 to balance the electrons, then adding and cancelling the electrons gives:

$2Li + 2H_2O \rightarrow 2Li^+ + 2OH^- + H_2$

So the required equation is $2Li + 2H_2O \rightarrow 2LiOH + H_2$.

Exam practice

1 Nitrogen dioxide reacts with water:

$2NO_2(g) + H_2O(l) \rightarrow H^+(aq) + NO_3^-(aq) + HNO_2(aq)$

(a) Give the oxidation states of nitrogen in all the nitrogen-containing species in the equation. [3]

(b) Which substance has been oxidised in the reaction? [1]

2 Scrap iron can be used to extract copper from dilute aqueous solutions containing copper(II) ions. Write the simplest ionic equation for the reaction of iron with copper(II) ions in aqueous solution. [1]

3 Ammonia reacts with oxygen in the presence of a platinum catalyst to form nitrogen(II) oxide and steam.

(a) Write a balanced equation for the reaction. [2]

(b) Identify which substances have been reduced and oxidised in the reaction. [2]

(c) What is the name of the reducing agent in the process? [1]

4 Which of the following shows chlorine in its correct oxidation states in the compounds shown? [1]

	HCl	$KClO_3$	HClO
A	−1	+3	+1
B	+1	−5	−1
C	−1	+5	+1
D	+1	+5	−1

Answers and quick quiz 7 online

ONLINE

Summary

You should now have an understanding of:

- what is meant by oxidation and reduction in terms of electron transfer
- what is meant by a redox reaction
- oxidation state
- how to deduce the oxidation state for an element in a compound
- how to use oxidation states to determine which elements have been oxidised and reduced in a reaction
- half-equations and how to write them for simple reactions
- how to combine half-equations to give ionic equations for redox reactions

8 Periodicity

Periodicity is the regular and **repeating pattern** of various physical or chemical properties. Since the outer electronic configuration of atoms is a periodic function, we therefore expect other properties to change accordingly.

Classification of elements in s, p, d or f blocks

The periodic table

REVISED

The periodic table is an arrangement of the known chemical elements according to their atomic numbers.

Elements are put into vertical columns called **groups**, in which all elements have similar chemical properties, as well as horizontal rows called **periods**.

Elements are classified as either s-, p-, d- or f-block elements according to their position in the periodic table. For example, Group 1 and 2 elements are in the s block (Figure 8.1) and their outer, highest-energy electrons occupy the *s*-sublevel.

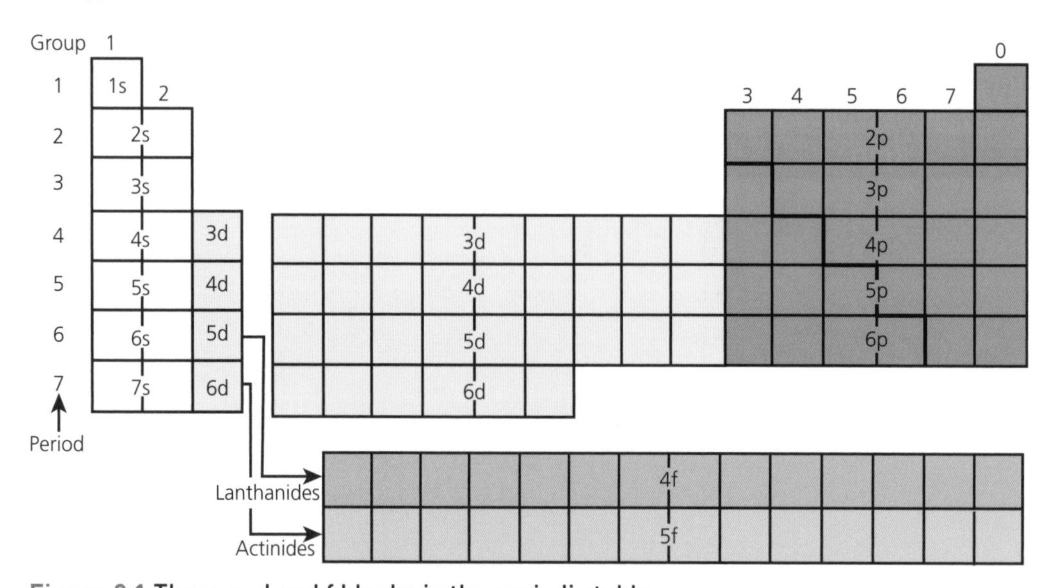

Figure 8.1 The s, p, d and f blocks in the periodic table

Now test yourself

TESTED

1 In which block of the periodic table are the elements in Groups 7 and 0 to be found?
2 The electronic configuration of boron, B, is $1s^2$, $2s^2$, $2p^1$. In which block of the periodic table would you expect boron be to be found?

Answers on p. 218

Properties of the Period 3 elements

Atomic radius

REVISED

Figure 8.2 shows atomic radius plotted against atomic number for Periods 2, 3 and 4 in the periodic table. A regular repeating pattern is seen; this is an example of periodicity.

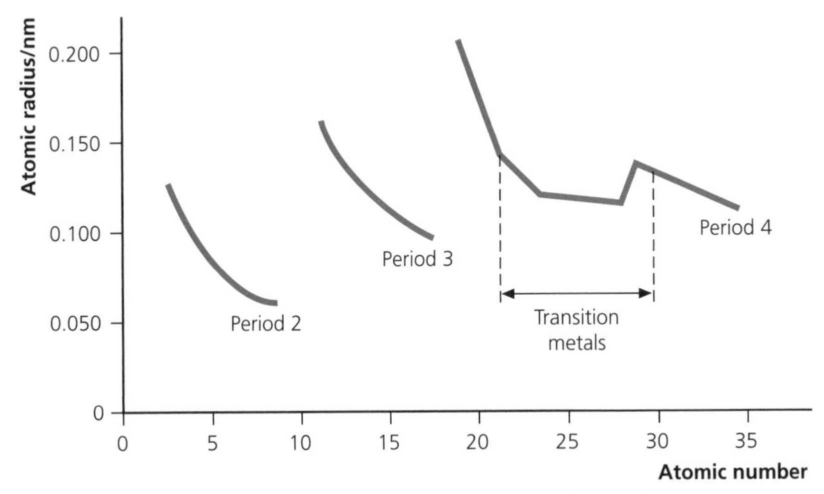

Figure 8.2 Graph of atomic radius versus atomic number

The graph shows two main trends.

Across a period

Moving from left to right across a period, the atomic radius generally decreases.

As the atomic number increases, the number of protons in the nucleus increases. Electrons are also being added, and to the same energy level. There is a greater electrostatic attraction between the outer electrons and the nucleus and so the atomic radius decreases.

Down a group

Electrons are being added to a higher energy level, progressively further from the nucleus. There is a reduction in the electrostatic attraction because of the increasing distance and so the atomic radius increases.

First ionisation energy

REVISED

Figure 1.4 on page 13 shows how the first ionisation energy changes with atomic number. Chapter 1 outlines the evidence for electron energy levels and sublevels, and gives an explanation of the ionisation energy trends observed down the groups and across a period.

Melting and boiling points

REVISED

On moving across the periodic table from left to right along Period 3, there is a gradual change in structure of the solid element (Table 8.1). When explaining the trends in melting and boiling points, it is important to know how the structures change.

Table 8.1 Element structures in Period 3

Group	1	2	3	4	5	6	7	0
Element	Na	Mg	Al	Si	P	S	Cl	Ar
Structural type	Giant metallic			Giant covalent	Simple covalent			Individual atoms

Figure 8.3 shows how the general trend seen in the melting points of the elements across a period relates closely to the trend seen in structure.

Revision activity

Using the internet, find the boiling points of the Period 2 elements (lithium to neon) and then plot them on a graph to see that the same pattern is seen as with the melting points.

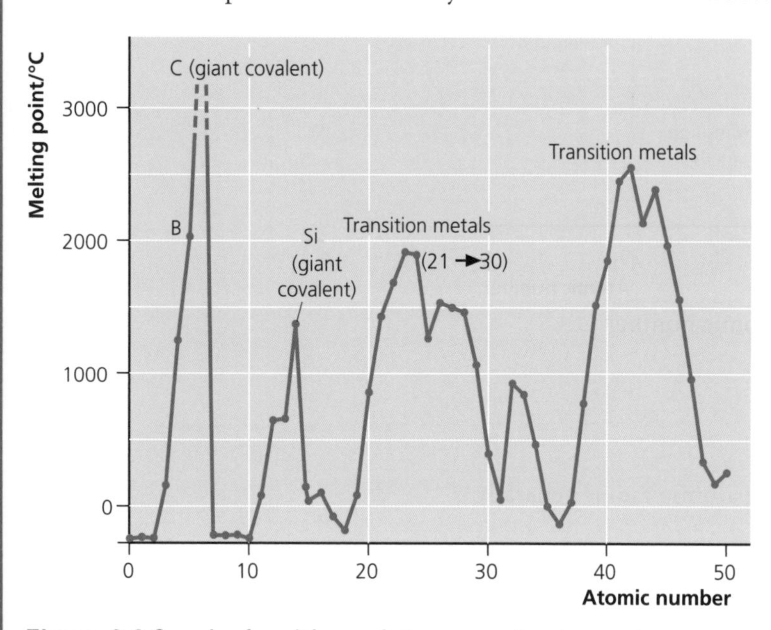

Figure 8.3 Graph of melting point versus atomic number

- The melting and boiling points of metals increase from Group 1 through to Group 3. This is because there are more electrons being delocalised into the **giant metallic structure**, and so there will be a greater electrostatic attraction between the positive ions and the mobile electrons. For example, the attraction between Al^{3+} ions and three electrons (on average) will be greater than that between Na^{+} ions and one electron. It will therefore be more difficult to separate the ions from each other in the melting or boiling process.
- The Group 4 element has the highest melting and boiling point in Period 3. Silicon atoms are strongly covalently bonded throughout in a **giant covalent structure** similar to that of diamond, so it will be difficult to melt or boil the element because this will involve breaking strong covalent bonds.
- There is a large drop after silicon but from then to the end of the period the melting point depends only on van der Waals forces, which in turn depend on the size of the molecules involved. In Group 5, phosphorus consists of separate P_4 molecules in a **simple covalent structure**. These molecules will be relatively easy to separate because their weak van der Waals forces will be easy to overcome. Sulfur in Group 6 consists of S_8 molecules, chlorine in Group 7 consists of Cl_2

molecules, and argon in Group 0 consists of separate atoms. Larger molecules with more electrons, for example S_8, give rise to larger van der Waals forces and hence their melting and boiling points are higher because it requires more energy to separate the molecules from each other.

Now test yourself

TESTED

3 Explain the following:
(a) Why does sodium have a larger atomic radius than magnesium?
(b) Why does aluminium have a lower first ionisation energy than magnesium?
(c) Why does sulfur have a higher boiling point than chlorine?

Answer on p. 218

Exam practice

1 An element X has the following first six ionisation energies in kJ mol^{-1}:
577, 1820, 2740, 11 600, 14 800, 18 400
(a) Explain how you know that element X is in Group 3 of the periodic table. [1]
(b) Element Y is in the same group as element X, but it is placed in the period below X in the periodic table.
(i) Give an approximate value for the first ionisation energy of element Y. [1]
(ii) Explain, using ideas of electronic structure, why you expect element Y to have this ionisation energy. [2]
(c) Two elements W and Z are in the same period as X, but W is in the group before X, and Z is in the group after X in the periodic table.
(i) Give approximate first ionisation energies for elements W and Z. [2]
(ii) Explain, using ideas of electronic structure, why elements W and Z have these ionisation energies. [2]

2 Indicate the general trend in the following properties across a period from Group 1 to Group 8:
(a) (i) first ionisation energy [1]
(ii) atomic radius [1]
(b) Explain the variation in melting points across a period. [2]

3 Explain these observations:
(a) Sodium has a larger atomic radius than lithium. [2]
(b) Silicon has a higher melting temperature than phosphorus, P_4. [2]

4 Which of these atoms has the highest electronegativity? [1]
A Na
B Mg
C Cl
D Ar

5 Which of these atoms has the largest atomic radius? [1]
A Ar
B Cl
C Mg
D Na

Answers and quick quiz 8 online

ONLINE

Summary

You should now have an understanding of:
- what is meant by the term 'periodicity'
- how elements can be classified in s, p, d or f blocks in the periodic table
- how atomic radius, first ionisation energy and melting and boiling point can be used to demonstrate periodicity

9 The periodic table

Group 2: the alkaline earth metals

The elements in Group 2 — beryllium, magnesium, calcium, strontium, barium and radium — are all metals.

Trends in physical properties

REVISED

Atomic radius

The atomic radius *increases* down the group, as shown in Figure 9.1.

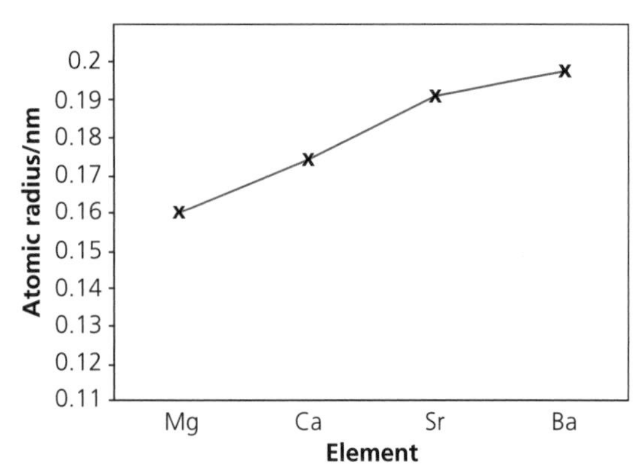

Figure 9.1 Atomic radii of Group 2 elements

- The number of electron energy levels increases down the group.
- The outer electrons experience more shielding.
- There will be a weaker electrostatic attraction between the outer electrons and the nucleus.
- So the atomic radius will increase.

Melting points of the elements

As Figure 9.2 shows, the general trend in melting point is for this to *decrease* down the group.

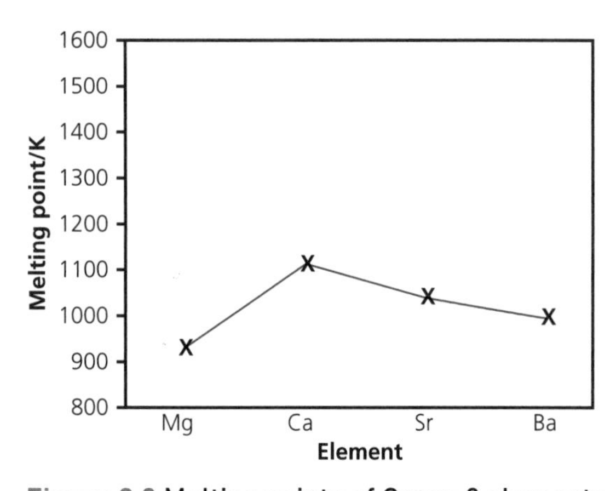

Figure 9.2 Melting points of Group 2 elements

Magnesium's melting point does not fit the general trend because it has a different metallic structure from the other elements in the group.

- All the elements are metals, so the structure of all these elements is **giant metallic**.
- Each atom donates two electrons to the delocalised electrons in the giant metallic structure.
- Ions of charge +2 form in the structure and these are bonded electrostatically by the mobile electrons.
- As the group is descended, the radius of the +2 ion increases because there are more electron energy levels.
- The outer electrons are further from the nucleus of each ion, and therefore further from the delocalised mobile electrons.
- The strength of the metallic bonding therefore decreases, and the melting point falls.

First ionisation energy of the elements

As Figure 9.3 shows, the first ionisation energy *decreases* down the group.

The **first ionisation energy** is the energy required to remove 1 mol of electrons from 1 mol of gaseous atoms under standard conditions:

$$M(g) \rightarrow M^{+}(g) + e^{-}$$

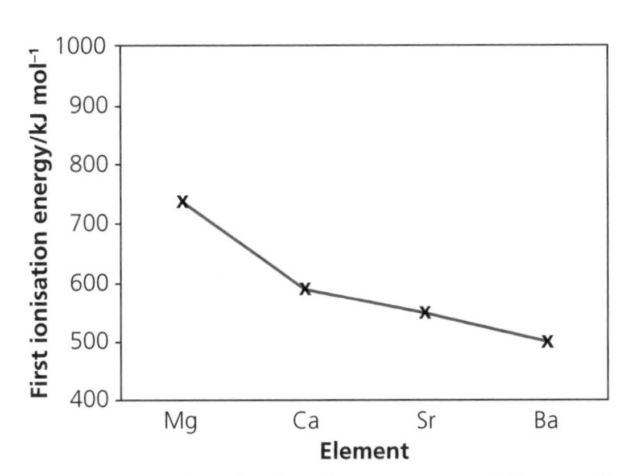

Figure 9.3 First ionisation energy of Group 2 elements

- There are more electron energy levels as the group is descended.
- The outer electron being removed is progressively further from the nucleus
- There will be less electrostatic attraction between the outer electron being removed and the nucleus.
- Therefore, the first ionisation energy will decrease.

Reactions of Group 2 elements

REVISED

All the metals in Group 2 are reactive, but not as reactive as Group 1 metals. They all react by losing their outer two *s*-electrons to form a positive ion (a cation) with a +2 charge.

Reaction with water

Group 2 metals react with cold water according to the general equation (where M is the Group 2 metal):

$$M(s) + 2H_2O(l) \rightarrow M(OH)_2(aq) + H_2(g)$$

A solution of the metal hydroxide (a weak base) forms along with hydrogen gas.

Reactivity *increases* down the group as it becomes progressively easier to remove two electrons because both the distance and the shielding between the outer electrons and the nucleus increase.

So the reactivity order for the metals, based on their rate of reaction with water is:

Mg ≪ Ca < Sr < Ba

Example

Write equations for the reactions of barium and magnesium with cold water. Comment on any differences in the observed rates of reaction.

Answer

- barium: $Ba(s) + 2H_2O(l) \rightarrow Ba(OH)_2(aq) + H_2(g)$
- magnesium: $Mg(s) + 2H_2O(l) \rightarrow Mg(OH)_2(aq) + H_2(g)$

The reaction of magnesium with water is much slower than that of barium. In fact, magnesium reacts only very slowly with cold water.

- The metal hydroxide that forms becomes more soluble as the group is descended.
- Because the hydroxide is more soluble as the group is descended, the concentration of hydroxide ions in solution will also be higher. The alkalis that are formed are stronger.
- The order of the relative solubility of the metal hydroxides and their strength as soluble bases (alkalis) is:

  $Mg(OH)_2 < Ca(OH)_2 < Sr(OH)_2 < Ba(OH)_2$

- Magnesium hydroxide is used as an antacid in medicine — it reacts with excess acid in the stomach and relieves indigestion. Calcium hydroxide is used in agriculture because it neutralises excess acidity in soil. Both uses exploit the compound's ability to react as a base.

Group 2 metals also react with steam, in which the reaction rate is considerably faster than that with cold water. Magnesium reacts violently with steam, as shown below:

$Mg(s) + H_2O(g) \rightarrow MgO(s) + H_2(g)$

Group 2 sulfates

Down the group, the sulfates become less soluble. The order of solubility is:

$MgSO_4 > CaSO_4 > SrSO_4 > BaSO_4$

When magnesium sulfate is added to water, a colourless solution is formed. When barium sulfate is added to water, almost none dissolves because it is extremely insoluble.

The **test for sulfate(VI) ions**, SO_4^{2-}, is to add hydrochloric acid and barium chloride solution. A white precipitate of barium sulfate(VI) will result. Hydrochloric acid is added to decompose any carbonate that may be present. A carbonate will also form a white precipitate with barium chloride solution and confuse the observations being made.

- Barium sulfate(VI), $BaSO_4$, is an insoluble compound and is easy to observe when **testing for a sulfate** because it forms a thick white precipitate.
- If sodium sulfate(VI) is the substance being tested, the reaction will be:

  $Na_2SO_4(aq) + BaCl_2(aq) \rightarrow BaSO_4(s) + 2NaCl(aq)$

  The **ionic equation** will be:

  $Ba^{2+}(aq) + SO_4^{2-}(aq) \rightarrow BaSO_4(s)$

- Barium sulfate is used in medicine when examining a patient's digestive system. The patient drinks a chalky barium sulfate suspension. When the patient is X-rayed, the barium sulfate coating inside the digestive tract absorbs a large proportion of the radiation. This highlights the black-and-white contrast of the X-ray photograph, so that doctors can diagnose digestive problems better.

Now test yourself

TESTED

1 Write an equation to show how calcium reacts with water. What would be observed in this reaction?
2 Describe a chemical test that would enable you to distinguish between sodium nitrate and sodium sulfate.

Answers on p. 218

Group 7: the halogens

- The halogens are fluorine (F), chlorine (Cl), iodine (I), bromine (Br) and astatine (At) — they are members of Group 7 of the periodic table.
- All halogens exist as diatomic molecules at room temperature and pressure, and are often given the general symbol X_2.

Trends in physical properties

REVISED

Boiling point

Boiling points *increase* down the group.

- The number of electrons in each molecule is increasing.
- This increases the number and strength of the **van der Waals interactions** between molecules.
- So it becomes harder to separate one molecule from another.

At room temperature fluorine is a pale-yellow gas, chlorine is a pale-green gas, bromine is an orange liquid and iodine is a dark-grey solid. Figure 9.4 shows clearly that the melting points and boiling points increase regularly down the group.

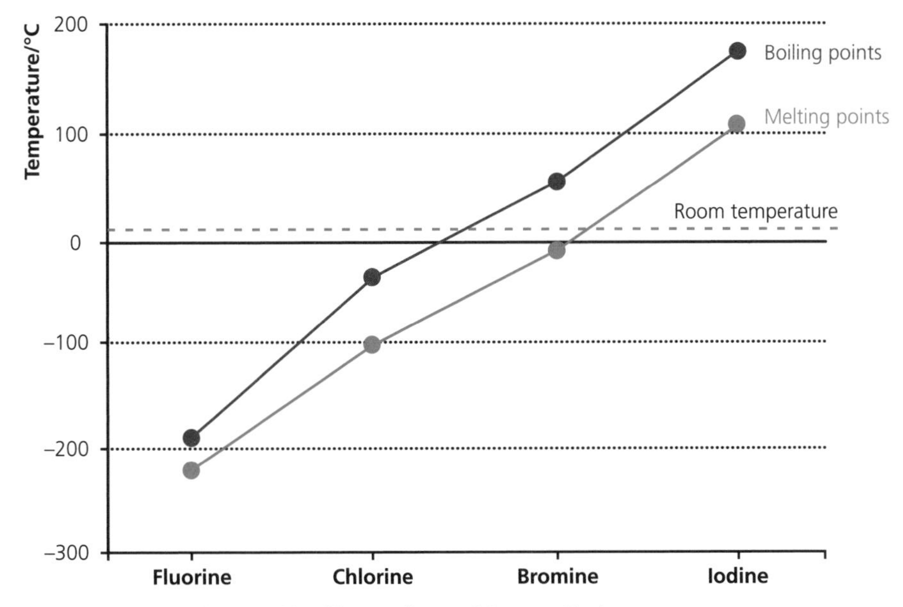

Figure 9.4 Melting and boiling points of Group 7 elements

Electronegativity

Electronegativity *decreases* down the group.

- The atomic radius increases because of the added electron shells.
- Therefore the bonding pair of electrons in the covalent bond gets progressively further from the positively charged nucleus. There will also be additional shielding.
- This means that there will be less electrostatic attraction between the nucleus and the bonding electron pair as the group is descended, and the electronegativity decreases.

Exam tip

Remember that the term 'electronegativity' refers to the ability of an atom to attract pairs of **bonding** electrons in covalent bonds.

Trends in the oxidising ability of halogens

REVISED

Halogen molecules, X_2, react by gaining electrons from other substances. A half-equation to show this process would be $X_2 + 2e^- \rightarrow 2X^-$.

In this process, the halogen molecule has been **reduced** because it has gained electrons. The other substance has been **oxidised** because it has had electrons removed by the halogen. This is why halogens are **oxidising agents**.

For example, sodium reacts with chlorine according to:

$$2Na(s) + Cl_2(g) \rightarrow 2NaCl(s)$$

Chlorine oxidises the sodium from oxidation state 0 to +1. Chlorine itself is reduced and changes oxidation number from 0 to −1.

The **oxidising power** of the halogens *decreases* down the group.

- As the group is descended, the number of electron energy levels increases.
- The distance between the outer electron shell and the nucleus (the atomic radius) increases.
- Shielding increases between the outer shell electrons and the nucleus.
- Because of the extra distance and shielding, the larger atoms attract an extra electron less strongly.
- Hence the oxidising power decreases down the group, and also the ability of a halogen molecule to be reduced.

Halogen–halide displacement reactions

A more reactive halogen will displace a less reactive halogen from its compounds.

When chlorine gas, $Cl_2(g)$, is bubbled through a solution containing sodium bromide, NaBr(aq), a **redox reaction** occurs in which the chlorine **oxidises** the bromide ions to aqueous bromine, $Br_2(aq)$.

- The overall equation for the reaction is $2NaBr(aq) + Cl_2(g) \rightarrow 2NaCl(aq) + Br_2(aq)$.
- The half-equation to show bromide ions being oxidised is $2Br^- \rightarrow Br_2 + 2e^-$.
- The half-equation to show chlorine molecules being reduced is $Cl_2 + 2e^- \rightarrow 2Cl^-$.
- The order of oxidising power of halogens is: $F_2 > Cl_2 > Br_2 > I_2$ (ignoring astatine).
- The order of reducing power of halide ions is: $I^- > Br^- > Cl^- > F^-$.
- Fluorine is a powerful oxidising agent and can oxidise Cl^-, Br^- and I^-.
- Chlorine can oxidise Br^- and I^-. Bromine can only oxidise I^-.

Exam tip

Make sure that you use the correct words — 'chlorine' for the molecular element and 'chloride' for the ion.

Now test yourself

TESTED

3 Explain why chlorine is a more powerful oxidising agent than iodine.
4 (a) Give an equation that shows how bromine reacts with sodium iodide solution.
(b) Which substance is oxidised and which is reduced in this reaction?

Answers on p. 218

Trends in the reducing ability of halide ions

REVISED

Halide ions increase in **reducing power** as the group is descended.

- There are more electron energy levels as the group is descended.
- The **ionic radius** of the halide ion increases.
- The amount of shielding between the outer electrons and the nucleus increases.
- There is a lower electrostatic attraction between the nucleus and the electrons.
- So it is easier to remove an electron from a halide ion as the group is descended.
- Therefore, **halide ions** are more readily oxidised — they are more powerful reducing agents.

The changing reducing ability of the halide ions can be demonstrated using a reaction between a halide salt and concentrated sulfuric acid.

Using NaCl(s):

sodium chloride + concentrated sulfuric acid → sodium hydrogen sulfate(VI) + hydrogen chloride gas

$$NaCl(s) + H_2SO_4(l) \rightarrow NaHSO_4(s) + HCl(g)$$

Using NaBr(s), and ignoring the Na^+ spectator ion:

$$Br^- + H_2SO_4 \rightarrow HSO_4^- + HBr$$

then:

$$2HBr + H_2SO_4 \rightarrow Br_2 + SO_2 + 2H_2O$$

Using NaI(s):

$$I^- + H_2SO_4 \rightarrow HSO_4^- + HI$$

then:

$$8HI + H_2SO_4 \rightarrow 4I_2 + H_2S + 4H_2O$$

- In the first stage, the halide ion is protonated by the concentrated sulfuric acid:

$$X^- + H^+ \rightarrow HX$$

- If a bromide or an iodide is used, a second stage occurs in which the sulfuric acid is reduced.
- The oxidation state of sulfur in sulfuric acid is +6. This is reduced to +4 in SO_2 and −2 in H_2S.
- The iodide ion is the most powerful reducing agent of the halide ions and produces a mixture of products including SO_2, S and H_2S in the reaction with sulfuric acid.

Typical mistake

Many students write an equation showing the formation of sodium sulfate(VI), Na_2SO_4. It is sodium *hydrogen*sulfate(VI) that forms when concentrated sulfuric acid is used.

Now test yourself

TESTED

5 Write an equation to show how potassium fluoride reacts with concentrated sulfuric acid.

6 (a) What are the different oxidation states of sulfur in H_2S, SO_2 and S?

(b) Write an equation for the reaction between hydrogen iodide and concentrated sulfuric acid to form sulfur, iodine and water.

Answers on p. 218

Identification of halide ions using silver nitrate

REVISED

Halide ions, X^-, can be identified using silver nitrate solution acidified with nitric acid. Nitric acid is used in the halide ion test because it reacts with any carbonates that may be present before silver nitrate is added. Silver ions react with carbonate ions to form white silver carbonate, and this will confuse the test.

Example

Using calcium chloride, an addition of acidified silver nitrate solution produces a white precipitate of silver chloride:

$$CaCl_2(aq) + 2AgNO_3(aq) \rightarrow Ca(NO_3)_2(aq) + 2AgCl(s)$$

Write an ionic equation for the reaction.

Answer

$Ag^+(aq) + Cl^-(aq) \rightarrow AgCl(s)$

The silver halide precipitates have similar colours and can be difficult to tell apart. A further test using aqueous ammonia of different concentrations can distinguish between them.

Table 9.1 Distinguishing silver halides

Silver halide formed	Colour of precipitate	Solubility in dilute $NH_3(aq)$	Solubility in concentrated $NH_3(aq)$
Silver chloride, AgCl(s)	White	Soluble — a colourless solution forms	Soluble — a colourless solution forms
Silver bromide, AgBr(s)	Cream	Insoluble	Soluble — a colourless solution forms
Silver iodide, AgI(s)	Yellow	Insoluble	Insoluble

Example

A solution of a substance Y is acidified with nitric acid and then a few drops of aqueous silver nitrate are added. A cream precipitate is formed. The precipitate is insoluble in dilute ammonia solution and does dissolve in concentrated ammonia.

What can you deduce from this information?

Answer

You can deduce that bromide ions must have been present in the compound, because the precipitate was cream in colour and dissolved in concentrated ammonia, but not in dilute ammonia.

Revision activity

Prepare some flashcards, on one side writing the type of halogen reaction (for example 'displacement with halide ions') and on the reverse writing the chemical equations that you must know.

Uses of chlorine and chlorate(I) compounds

REVISED

Reaction of chlorine with water

Chlorine reacts with water according to the equation:

chlorine + water → chloric(I) acid + hydrochloric acid

$Cl_2(g) + H_2O(l) \rightarrow HOCl(aq) + HCl(aq)$

- The chloric(I) acid formed, $HOCl(aq)$, is a powerful oxidising agent. Chlorine is in oxidation state +1.
- In the above reaction, the chlorine initially is in oxidation state 0. At the end of the reaction, chlorine is in oxidation state +1 in chloric(I) acid (an oxidation); in hydrochloric acid it is in oxidation state −1 (a reduction).
- The powerful oxidising nature of chloric(I) acid is exploited in using chlorine to purify water supplies for drinking.
- Although chlorine is poisonous, its benefits in killing microorganisms responsible for water-borne diseases, such as cholera, typhoid and diphtheria, outweigh any potential disadvantages.
- Chloric(I) acid, $HOCl(aq)$, is a bleaching and oxidising agent, and it decomposes in the presence of sunlight to form oxygen:

$2HOCl(aq) \rightarrow 2HCl(aq) + O_2(g)$

Typical mistake

Many students know that chlorine is added to water to disinfect it; but few understand that its action is due to the oxidising nature of the chloric(I) acid formed.

Reaction of chlorine with cold, aqueous, dilute sodium hydroxide solution

Chlorine disproportionates in cold, dilute aqueous alkali:

$Cl_2(g) + 2NaOH(aq) \rightarrow NaOCl(aq) + NaCl(aq) + H_2O(l)$

- Chlorine's oxidation state increases from 0 to +1 and also decreases from 0 to −1. The product of this reaction, sodium chlorate(I) ($NaOCl$), is used as a **bleach** and as a **disinfectant** to kill germs.
- The chlorate(I) ion **oxidises** germs and bacteria by removing electrons:

$ClO^-(aq) + 2H^+(aq) + 2e^- \rightarrow Cl^-(aq) + H_2O(l)$

Now test yourself

TESTED

7 Describe a chemical test to distinguish between lithium chloride and lithium iodide. Write ionic equations for the reactions that take place.

8 (a) Give different uses for chlorine and sodium chlorate(I).

(b) Give reagents for a reaction in which chloric(I) acid forms.

Answers on p. 218

Exam practice

1 Concentrated sulfuric acid is added separately to potassium chloride and potassium iodide.
   (a) Write equations to show each of the reactions. [2]
   (b) Predict two expected observations when potassium astatide (KAt) is added to concentrated sulfuric acid. [2]
   (c) Explain these trends down Group 7 using the electronic structures of the atoms to help you:
      (i) electronegativity [2]
      (ii) boiling point [2]
      (iii) atomic radius [2]
   (d) Write equations for these reactions involving halogens or halide ions:
      (i) potassium with fluorine gas [2]
      (ii) sodium bromide solution with chlorine gas [2]
      (iii) chlorine gas with water [2]
      (iv) sodium chloride solution with silver nitrate solution [2]

2 (a) Give uses for these substances: [4]
      (i) calcium hydroxide, $Ca(OH)_2$
      (ii) magnesium hydroxide, $Mg(OH)_2$
      (iii) chlorine, Cl_2
      (iv) barium sulfate(VI), $BaSO_4$
   (b) Describe chemical tests to distinguish between each of the compounds in these pairs:
      (i) potassium iodide and potassium bromide [3]
      (ii) copper(II) chloride and copper(II) sulfate [3]

3 State and explain the trend in the first ionisation energies of the elements in Group 2 from magnesium to barium. [3]

4 Which of these elements has the highest second ionisation energy? [1]
   A Na
   B Mg
   C Ne
   D Ar

Answers and quick quiz 9 online

ONLINE

Summary

You should now have an understanding of:

- how atomic radius, melting point and first ionisation energy vary down Group 2 of the periodic table
- reactions of Group 2 metals with water
- the variation in the solubility of Group 2 hydroxides
- how to test for sulfate(VI) ions using acidified barium chloride solution
- the variation in solubility of Group 2 sulfates
- how electronegativity, melting point and boiling point vary down Group 7 of the periodic table
- the oxidising power of the halogens
- halogen–halide displacement reactions
- the relative reducing power of the halides shown in their reactions with halogens and with concentrated sulfuric acid
- how to test for a halide ion using acidified silver nitrate solution
- how chlorine reacts with water and cold, dilute aqueous sodium hydroxide

10 Alkanes

Fractional distillation of crude oil

Alkanes are **hydrocarbons** — this means that their molecules contain *only* hydrogen and carbon atoms. The general formula of an alkane is C_nH_{2n+2}.

Alkanes are **saturated** — this means that they contain only carbon–carbon single bonds — no multiple bonds.

The alkanes are found in crude oil and are separated from this by a process called **fractional distillation**. Crude oil is a complex mixture of mainly hydrocarbons. The complex mixture can be separated into smaller mixtures or fractions that have different boiling points. These different components of the mixture can be drawn off at different levels in a fractionating column because of the temperature gradient up the column (Figure 10.1).

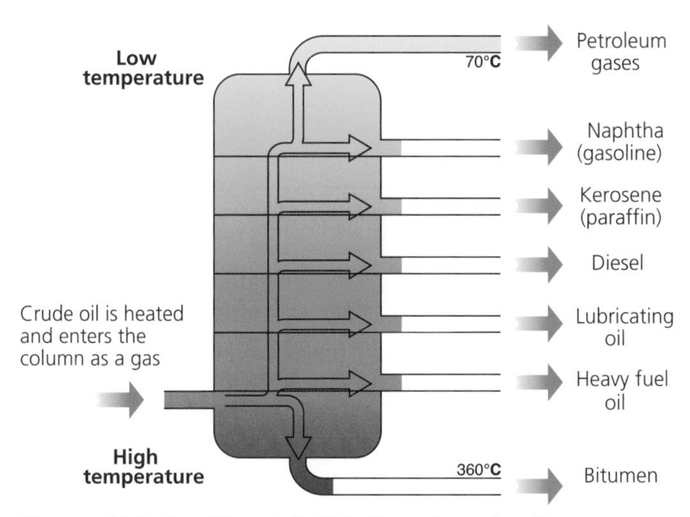

Figure 10.1 Fractional distillation of crude oil

Formulae and naming

Formulae

REVISED

The first five members of the alkane **homologous series** are shown in Figure 10.2.

```
    H            H  H              H  H  H
    |            |  |              |  |  |
H — C — H    H — C — C — H     H — C — C — C — H
    |            |  |              |  |  |
    H            H  H              H  H  H
methane        ethane            propane

    H  H  H  H               H  H  H  H  H
    |  |  |  |               |  |  |  |  |
H — C — C — C — C — H    H — C — C — C — C — C — H
    |  |  |  |               |  |  |  |  |
    H  H  H  H               H  H  H  H  H
      butane                   pentane
```

Figure 10.2 The first five alkanes

Because they form a homologous series:

- they have the same general formula
- they contain the same functional groups and have similar chemical properties
- each member differs from the next by a CH_2 unit

The **molecular formula** gives the atoms of each element in a molecule. The **empirical formula** gives the simplest whole number ratio of atoms of each element in a molecule.

Using butane as an example:

- the **displayed formula** showing all the bonds is

```
    H   H   H   H
    |   |   |   |
H — C — C — C — C — H
    |   |   |   |
    H   H   H   H
```

- the **structural formula** is $CH_3CH_2CH_2CH_3$
- the **molecular formula** is C_4H_{10} — this shows the actual number of atoms of each element in the molecule, but shows no structure
- the **empirical formula** is C_2H_5 — this shows the simplest whole-number ratio for the numbers 4 and 10 in the molecular formula

Now test yourself

TESTED

1 Write the displayed formula, structural formula, molecular formula and empirical formula for *n*-hexane, C_6H_{14}.

Answer on p. 218

Exam tip

Ensure that you understand the differences between the types of formula — they can sometimes be very different.

Naming alkanes

REVISED

Unbranched alkanes are easy to name because they follow the 1: meth-, 2: eth-, 3: prop-, 4: but- system with -ane at the end of the name. For example, the alkane containing a longest chain of three carbon atoms is called propane.

Branched-chain alkanes are named according to the following steps:

1. Find the longest continuous chain of carbon atoms.
2. Look for any branches along the longest chain — how many carbons atoms are there in each branch and at what positions are they along the longest chain?

Exam tip

The arrangement around each carbon atom is actually tetrahedral. When the structures are 'squashed' onto a flat page, the angles look like 90° or 180°, but they are all 109.5°.

Example

Name the molecules in Figure 10.3.

(a)

```
    H   H    H   H
    |   |    |   |
H — C — C —  C — C — H
    |   |    |   |
    H   CH3  H   H
```

(b)

```
    H   CH3  H   H   H   H
    |   |    |   |   |   |
H — C — C —  C — C — C — C — H
    |   |    |   |   |   |
    H   CH3  H   H   H   H
```

Figure 10.3

Answer

(a) It has four carbon atoms in its longest chain.
It has one methyl group positioned at the number 2 carbon atom (counting from the shortest end).
Name: 2-methylbutane

(b) It has six carbon atoms in its longest chain.
It has two methyl groups and they are both at the number 2 position.
Name: 2,2-dimethylhexane

Now test yourself

TESTED

2 Name the alkanes in Figure 10.4, all of which are isomers of C_5H_{12}.

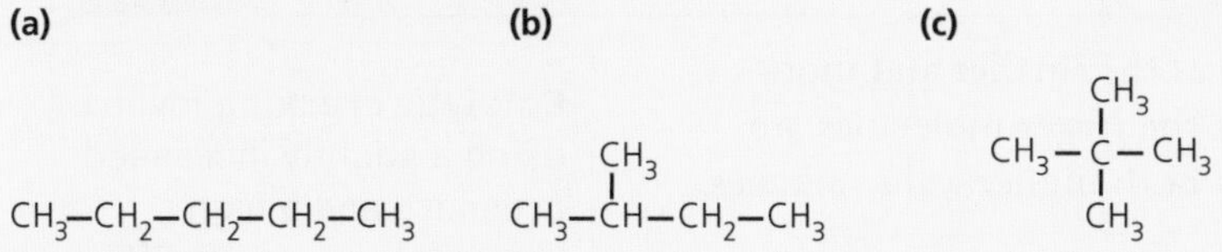

(a) $CH_3-CH_2-CH_2-CH_2-CH_3$

(b) $CH_3-CH(CH_3)-CH_2-CH_3$

(c) $CH_3-C(CH_3)_2-CH_3$

Figure 10.4

Answers on p. 218

Isomerism

Structural isomers

REVISED

Many organic molecules, like alkanes, can form more than one structure with the same number of atoms in each molecule — these are called **isomers**.

There are several different types of structural isomerism — **chain**, **position** and **functional group**.

Structural isomers are molecules with the same molecular formula but different structural formulae.

Chain isomerism

REVISED

These isomers arise because of the possibility of branching in carbon chains. For example, there are two isomers of C_4H_{10}. In one of them, the carbon atoms lie in a 'straight' chain whereas in the other the chain is branched (Figure 10.5).

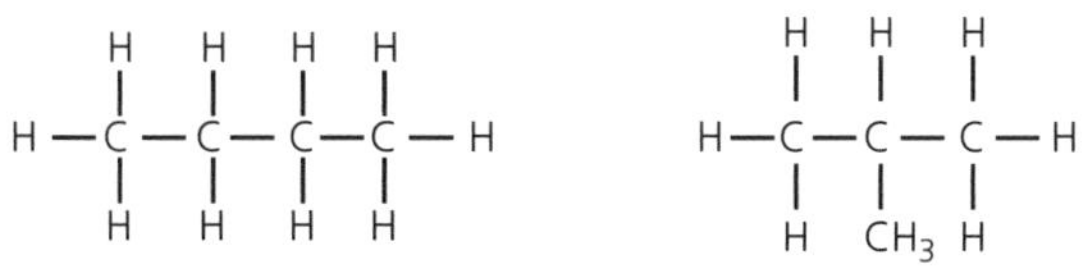

butane: $CH_3-CH_2-CH_2-CH_3$ (displayed formula)

2-methylpropane: $CH_3-CH(CH_3)-CH_3$ (displayed formula)

Figure 10.5

Position isomerism

REVISED

In position isomerism, the basic carbon skeleton remains unchanged, but important groups are attached to different carbon atoms along the chain (Figure 10.6).

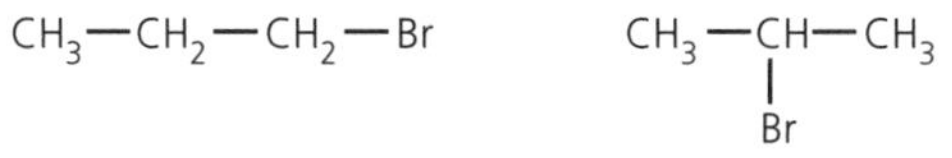

$CH_3-CH_2-CH_2-Br$ — 1-bromopropane

$CH_3-CH(Br)-CH_3$ — 2-bromopropane

Figure 10.6

Functional group isomerism

REVISED

In this type of structural isomerism, the isomers contain different functional groups — that is, they belong to different families of compounds (different homologous series). For example, a molecular formula C_3H_6O could be either propanal (an aldehyde) or propanone (a ketone) (Figure 10.7).

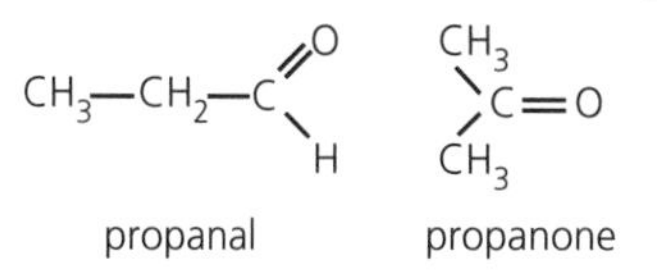

Figure 10.7

Cracking of alkanes

Breaking carbon–carbon bonds

REVISED

Longer-chain alkanes are not as marketable as the smaller and more useful alkanes. Because of this low demand, the larger molecules are **thermally** or **catalytically cracked** to form both higher-value alkanes and alkenes.

Cracking involves breaking carbon–carbon bonds in an alkane, and so energy is required to make this happen. It is the process by which long-chain alkanes are converted into smaller alkanes and, normally, alkenes.

For example, the alkane decane, $C_{10}H_{22}$, could be cracked as follows:

$$C_{10}H_{22}(g) \rightarrow C_8H_{18}(g) + C_2H_4(g)$$

Catalytic cracking involves using a slightly increased pressure and a high temperature — a **zeolite catalyst** is often used. Aromatic hydrocarbons and motor fuel are two of the important products formed in this catalytic cracking process.

Thermal cracking involves using a high temperature at high pressure. This process produces a high percentage of **alkenes**.

Now test yourself

TESTED

3 Write an equation to show the cracking of the alkane nonane, C_9H_{20}, to form ethene and one other product.

Answer on p. 218

Exam tip

Cracking often produces ethene, C_2H_4, which is used to make the polymer poly(ethene).

Combustion of alkanes

Fuels

REVISED

Alkanes are often used as fuels when burned in oxygen. The products formed depend on the amount of oxygen gas available. For example, propane gas, C_3H_8, can be burned in excess oxygen — complete combustion: $C_3H_8(g) + 5O_2(g) \rightarrow 3CO_2(g) + 4H_2O(l)$

Alternatively it can burn by partial or incomplete combustion, in which the oxygen supply is limited: $C_3H_8(g) + \frac{7}{2}O_2(g) \rightarrow 3CO(g) + 4H_2O(l)$

Exam tip

Note that water is always formed in hydrocarbon combustion. However, carbon monoxide and/or carbon can be formed when the oxygen supply is limited.

Now test yourself

TESTED

4 If 100 cm^3 of ethane is burned completely, what volume of oxygen is required to ensure that this happens? Assume that all volumes are measured under the same conditions.
5 Write an equation to show the incomplete combustion of methane in which gases are formed.

Answers on p. 218

Environmental issues

REVISED

In a typical car engine in which diesel or petrol is the fuel, carbon monoxide, unburned hydrocarbons and oxides of nitrogen (NO_x) gases are found in exhaust gases — all of these are potential pollutants and can present a problem to human health.

Pollutants such as nitrogen monoxide, NO(g), and carbon monoxide, CO(g), are removed from car exhaust fumes using a catalytic converter.

The gases react *together* on the surface of a catalyst, such as rhodium, palladium or platinum, to form products that are less harmful:

$$2NO(g) + 2CO(g) \rightarrow 2CO_2(g) + N_2(g)$$

If fuels contain sulfur, then sulfur dioxide will form when the fuel is burned. This gas is acidic and contributes to acid rain. It can be removed by passing it through powdered calcium oxide (a base) so that it reacts. An acid–base neutralisation occurs: $CaO(s) + SO_2(g) \rightarrow CaSO_3(s)$

Gases, such as carbon dioxide and water vapour, and unburned hydrocarbons, such as methane, are all **greenhouse gases**. This means that they are able to absorb infrared radiation, which results in the temperature of the atmosphere increasing. This is called **global warming**.

Now test yourself

TESTED

6 Give the names of two chemical pollutants that are formed when a fuel such as petrol is burned. How are these pollutants removed?

Answer on p. 218

Exam practice

1 The alkane pentane is a saturated hydrocarbon.
  (a) (i) State the meaning of the terms 'saturated' and 'hydrocarbon' as applied to alkanes. [2]
    (ii) Give the general formula for the alkanes. [1]
  (b) Pentane burns completely in oxygen.
    (i) Write an equation for this reaction. [1]
    (ii) State how the product of this reaction may affect the environment. [1]
  (c) Give the name of the solid pollutant that may form when pentane burns incompletely in air. [1]
  (d) A molecule of $C_{13}H_{28}$ can be cracked to propene and one other product.
    (i) Write an equation for this cracking reaction. [1]
    (ii) Suggest a type of compound that can be manufactured from the propene produced in this cracking reaction. [1]
    (iii) State why a high temperature is required for cracking reactions to occur. [1]

2 What is the total volume of gas remaining after 20 cm^3 of ethane is burned completely in 100 cm^3 oxygen? All volumes are measured at the same pressure and the same temperature, which is above 100°C. [1]

$C_2H_6 + 3\frac{1}{2}O_2 \rightarrow 2CO_2 + 3H_2O$

A 40 cm^3 B 100 cm^3 C 120 cm^3 D 130 cm^3

Answers and quick quiz 10 online

ONLINE

Summary

You should now have an understanding of:

- what the alkanes are
- the terms 'empirical formula', 'molecular formula', 'structural formula' and 'displayed formula'
- homologous series and functional groups
- how to name simple alkanes containing up to six carbons
- what is meant by 'isomers' and, in particular, the various types of structural isomer
- the fractional distillation of crude oil
- the cracking of long-chain alkanes and uses for the products
- the different types of combustion of alkanes
- the consequences of combustion in terms of pollution and how some pollutants are removed

11 Halogenoalkanes

Halogenoalkanes can be classed as **primary**, **secondary** or **tertiary** depending on how many alkyl groups (methyl-, ethyl-, propyl-) are attached to the carbon atom that is bonded to the halogen atom. The halogenoalkanes shown in Figure 11.1 are classed as primary, secondary and tertiary respectively.

Halogenoalkanes are organic molecules based on alkanes in which one or more of the hydrogen atoms have been replaced by halogen atoms.

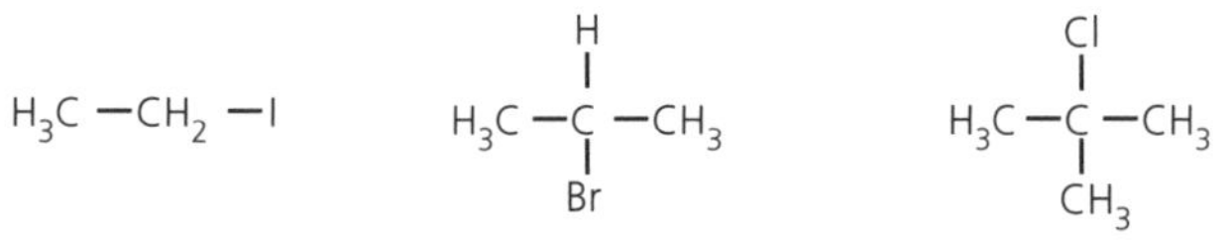

H_3C-CH_2-I iodoethane

$H_3C-CH(Br)-CH_3$ 2-bromopropane

$H_3C-C(Cl)(CH_3)-CH_3$ 2-chloro-2-methylpropane

Figure 11.1 Examples of halogenoalkanes

Synthesis of chloroalkanes

Chloroalkanes are compounds in which some hydrogen atoms of alkanes have been replaced by one or more chlorine atoms.

Methane with chlorine

REVISED

Methane, CH_4, reacts with chlorine, Cl_2, in the presence of ultraviolet light in a reaction **mechanism** called **free-radical substitution**.

The overall equation for the reaction is:

$$CH_4 + Cl_2 \rightarrow CH_3Cl + HCl$$

The mechanism occurs in three stages.

- **Initiation:** chlorine **radicals** are formed in the presence of ultraviolet light:

  $$Cl_2 \rightarrow 2Cl\bullet$$

- **Propagation:** chlorine radicals react with methane molecules to form new radicals and molecules. Adding the propagation steps together gives the overall equation for the reaction:

  $$CH_4 + Cl\bullet \rightarrow HCl + \bullet CH_3$$

  then:

  $$\bullet CH_3 + Cl_2 \rightarrow CH_3Cl + Cl\bullet$$

- **Termination:** radicals react with each other to form molecules:

  $$\bullet CH_3 + Cl\bullet \rightarrow CH_3Cl \text{ or } \bullet CH_3 + \bullet CH_3 \rightarrow C_2H_6$$

Radicals are molecules, atoms or ions that have an unpaired electron. This makes them very reactive.

Further substitutions, in which more hydrogen atoms are replaced by halogens, are possible. Using excess alkane will limit this happening.

Halogenoalkanes, like chloroalkanes and chlorofluoroalkanes (CFC), can be used as solvents.

Ozone depletion

Ozone gas, O_3, is formed in the upper atmosphere of our planet. It protects life on Earth from potentially harmful ultraviolet radiation from the Sun.

Chlorine radicals, Cl•, react with ozone resulting in its removal or depletion. Chlorine radicals are formed in the upper atmosphere from halogenoalkanes containing C–Cl bonds, for example the chlorofluorocarbon dichlorodifluoromethane, CCl_2F_2. The reaction involves a sequence of initiation and propagation steps.

- **Initiation:** formation of chlorine radicals from the CFC in the presence of ultraviolet light:

  $CCl_2F_2 \rightarrow \bullet CClF_2 + Cl\bullet$

- **Propagation:** reaction of chlorine radicals with ozone molecules:

  $Cl\bullet + O_3 \rightarrow ClO\bullet + O_2$

  then:

  $ClO\bullet + O_3 \rightarrow 2O_2 + Cl\bullet$

Adding these two propagation equations gives the overall equation for the reaction:

$2O_3 \rightarrow 3O_2$

Chlorine radicals catalyse the decomposition of ozone. Notice how they are used up in the reaction and then reproduced at the end, this being typical catalytic behaviour. This means that chlorine radicals can continue to destroy ozone for many years.

Hydrofluorocarbons, HFCs, and hydrochlorofluorocarbons, HCFCs, are now being used as alternatives to CFCs. However, HCFCs can still deplete the ozone layer. Although their depleting effect is only about one-tenth that of CFCs, serious damage is still being caused. HCFCs are a short-term fix until better replacements can be developed.

Now test yourself

TESTED

1 (a) Explain why CFCs are damaging to the ozone layer.
  (b) State what is meant by a 'radical'.
  (c) Give two equations to show how chlorine radicals attack ozone.

2 Ethane gas reacts with bromine in the presence of ultraviolet light. Bromoethane is one of the products of the reaction.
  (a) Write an overall equation for the main reaction taking place.
  (b) Name the mechanism by which the reaction takes place.
  (c) Outline the mechanism for the reaction.

Answers on p. 218

Nucleophilic substitution

Polar molecules

REVISED

Halogenoalkanes are **polar molecules**. They contain electronegative halogen atoms, X, that are attached to carbon atoms and the bond is polarised:

$^{\delta+}C{-}X^{\delta-}$

Halogenoalkanes can react by having their carbon–halogen bond broken. The ease of this breaking depends on the strength of the covalent bond. They can react with **nucleophiles** in which the halogen atom is replaced by a nucleophile.

A **nucleophile** is a donor of a lone pair of electrons.

In the case of **nucleophilic substitution**, a **nucleophile** attacks the slightly positively charged carbon atom. The halogen atom then is substituted by the nucleophile.

Common nucleophiles include: hydroxide, **:**OH^-; cyanide, **:**CN^- and ammonia, **:**NH_3.

The ease of carbon–halogen bond breaking is:

C–I > C–Br > C–Cl > C–F

Therefore, iodoalkanes react faster than other halogenoalkanes because the C–I bond is the weakest.

The order of reactivity is:

iodoalkanes > bromoalkanes > chloroalkanes > fluoroalkanes

Nucleophilic substitution reactions are highly important in organic synthesis because replacing a halogen atom with another functional group is useful and convenient.

Typical mistake

It is a common misconception for students to explain the rate of reaction of halogenoalkanes with nucleophiles in terms of C–X bond polarity, when it is in fact *bond strength* that is key.

Reaction with aqueous sodium hydroxide solution

REVISED

Halogenoalkanes undergo substitution reactions with hot, aqueous sodium hydroxide solution in which the **:**OH^- ion from the alkali acts as a **nucleophile** (lone-pair donor) and attacks the carbon atom attached to the halogen, which is lost as a halide ion (Figure 11.2). For example, using chloromethane:

$$CH_3Cl + OH^- \rightarrow CH_3OH + Cl^-$$

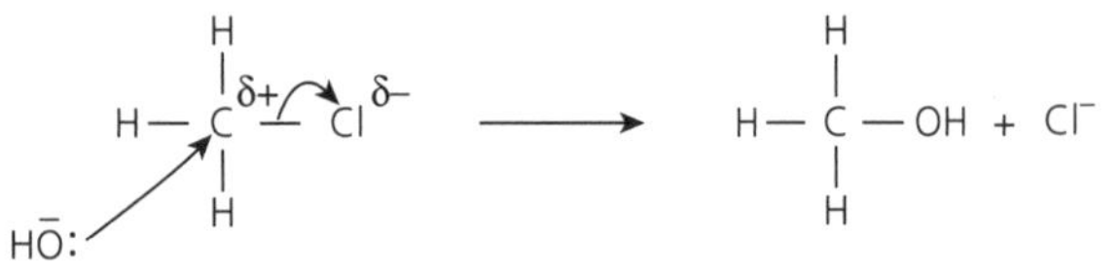

Figure 11.2 The nucleophilic substitution of chloromethane by hydroxide ions

In this reaction, the hydroxide ion nucleophile donates a lone pair of electrons to the carbon atom with the $\delta+$ charge in the halogenoalkane molecule. This is shown using a 'curly arrow' in the mechanism. The two bonding electrons in the carbon–halogen bond then move onto the halogen atom, and a halide ion forms, again shown by a curly arrow.

Reaction with sodium cyanide

REVISED

Cyanide ions, $:CN^-$, from sodium cyanide act as nucleophiles towards halogenoalkanes:

$$CH_3I + CN^- \rightarrow CH_3CN + I^-$$

The mechanism for this process (Figure 11.3) is identical to the previous cases using hydroxide ions, OH^-:

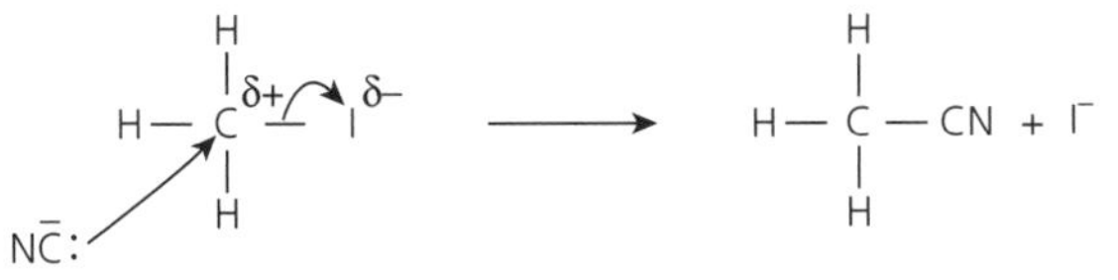

Figure 11.3 Cyanide ions reacting with iodomethane

The reaction with ethanolic ammonia

REVISED

An ammonia molecule acts as a nucleophile and donates its lone pair of electrons to the carbon atom attached to the halogen (Figure 11.4). For example:

$$C_2H_5Br + 2NH_3 \rightarrow C_2H_5NH_2 + NH_4Br$$

The ethylamine that forms is a base and reacts with hydrogen bromide (an acid) to form ammonium bromide. This reaction requires ethanol as a solvent and needs an excess of ammonia to limit further substitution.

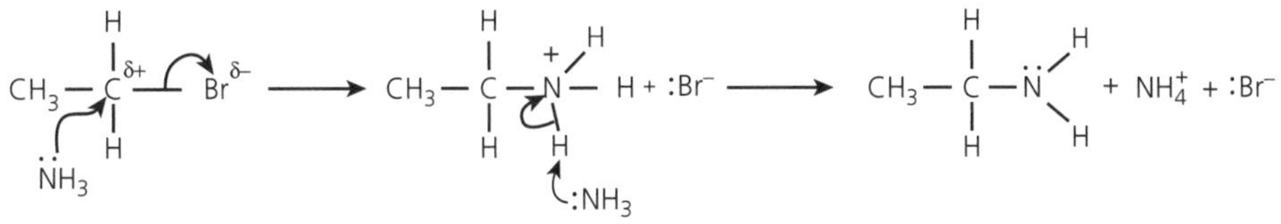

Figure 11.4 The nucleophilic reaction between ammonia and bromoethane

Elimination

Elimination reactions

REVISED

When potassium hydroxide solution, KOH(aq), is added to 2-bromopropane, a nucleophilic substitution reaction in which the hydroxide ion acts as a nucleophile can happen:

$$CH_3\text{-}CHBr\text{-}CH_3 + OH^- \rightarrow CH_3\text{-}CH(OH)\text{-}CH_3 + Br^-$$

However, it is also possible for the hydroxide ion to behave as a **base** — a proton is removed from one of the carbon atoms next to the carbon attached to the halogen atom. In this case, an **elimination reaction** takes place, forming an alkene (Figure 11.5).

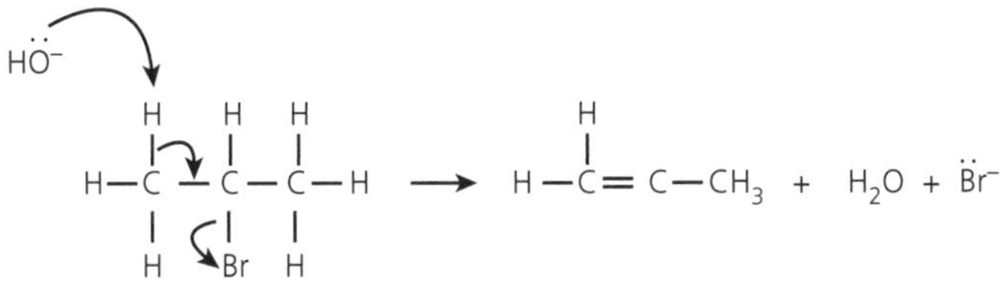

Figure 11.5 Mechanism to show an elimination process

Exam tip

The choices of solvent and temperature have a crucial role in this type of reaction. A hot ethanol solvent promotes elimination; warm aqueous conditions promote nucleophilic substitution.

The solvent, ethanol, promotes the elimination process over the nucleophilic substitution process.

The products of this reaction are propene, water and bromide ions:

$$CH_3\text{–}CHBr\text{–}CH_3 + {}^-OH \rightarrow CH_3\text{–}CH{=}CH_2 + Br^- + H_2O$$

Exam practice

1 (a) Name the halogenoalkanes in Figure 11.6. [4]

```
(i)                          (ii)                  (iii)                    (iv)

                              H   Br   H            H   H   H   H             H   H   H
  H   H   H   H               |   |    |            |   |   |   |             |   |   |
H-C - C - C - C-Br         H- C - C -  C -H       H-C - C - C - C-H        H- C - C - C -Br
  |   |   |   |               |   |    |            |   |   |   |             |   |   |
  H   H   H   H              H H-C- H H             H   H   Br  H            H H-C- H H
                                  |                                               |
                                  H                                               H
```

Figure 11.6

(b) Classify the halogenoalkanes in part (a) as primary, secondary or tertiary. [4]

(c) How are the four molecules related? [1]

(d) Draw structural formulae to show the products formed when 1-iodobutane reacts with:

(i) excess ammonia in ethanol [1]

(ii) warm NaOH(aq) [1]

(iii) NaCN [1]

(e) Draw the structure of the product in part (d)(ii) if hot ethanol were used as the solvent. [2]

(f) How would the rates of reaction differ when chloroethane and bromoethane are reacted separately with aqueous sodium hydroxide solution? Explain your answer. [2]

2 (a) CCl_4 is an effective fire extinguisher but it is no longer used because of its toxicity and its role in the depletion of the ozone layer. In the upper atmosphere, a bond in CCl_4 breaks and reactive species are formed. Identify the condition that causes a bond in CCl_4 to break in the upper atmosphere. Deduce an equation for the formation of the reactive species. [2]

(b) One of the reactive species formed from CCl_4 acts as a catalyst in the decomposition of ozone. Write two equations to show how this species acts as a catalyst. [2]

(c) A small amount of freon, CF_3Cl, with a mass of 1.78×10^{-4} kg escaped from a refrigerator into a room of volume $100\,m^3$. Assuming that the freon is evenly distributed throughout the air in the room, calculate the number of freon molecules in a volume of $500\,cm^3$. Give your answer to the appropriate number of significant figures. The Avogadro constant = $6.02 \times 10^{23}\,mol^{-1}$. [3]

3 Which of these substances reacts most rapidly to produce a silver halide precipitate with acidified silver nitrate? [1]

A CH_3Br

B CH_3Cl

C CH_3F

D CH_3I

Answers and quick quiz 11 online

ONLINE

Summary

You should now have an understanding of:

- how methane reacts with chlorine in the presence of ultraviolet light
- the free-radical substitution process
- how ozone is depleted by the presence of chlorine radicals
- nucleophilic substitution mechanisms
- how halogenoalkanes react with NaOH, KCN and excess NH_3
- how a hydroxide ion can behave either as a nucleophile or a base and how the products differ depending on the conditions used
- elimination processes involving halogenoalkanes

12 Alkenes

Structure, bonding and reactivity

Carbon–carbon double bonds

REVISED

Alkenes are hydrocarbons that contain at least one carbon–carbon double bond. For this reason they are described as being **unsaturated**.

The presence of a double bond makes alkenes *more reactive* than alkanes. This is because the double bond is an area of high electron density in the molecule.

The structure of the carbon–carbon double bond is shown in Figure 12.1.

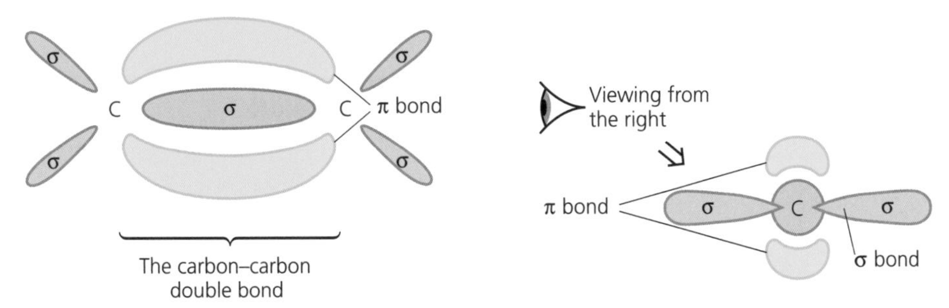

Figure 12.1 Representation of the σ and π bonds in a double bond

- The carbon–carbon double bond arrangement is **planar** (flat) and the internal bond angle is 120°.
- The presence of the π bond prevents rotation about the σ bond.
- This lack of rotation gives rise to ***E–Z* (geometrical) isomers**.
- *E–Z* geometrical isomerism is one example of **stereoisomerism**.

Stereoisomers are compounds with the same structural formula but with bonds arranged differently in space.

The *E–Z* isomers in Figure 12.2 show how molecules such as but-2-ene and 1,2-dichloroethene can exist in more than one form. A pair of *E–Z* isomers differ because of the different arrangement of groups in space.

H, CH_3 / C=C / CH_3, H

E-but-2-ene

H_3C, CH_3 / C=C / H, H

Z-but-2-ene

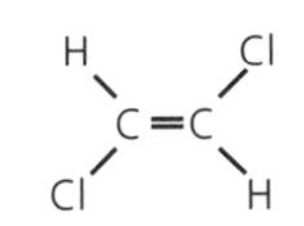

E-1,2-dichloroethene

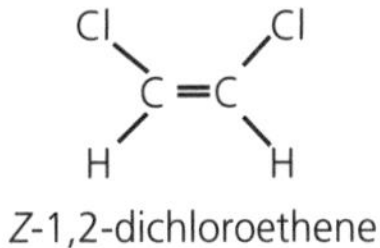

Z-1,2-dichloroethene

Figure 12.2 Geometric isomers

The letter '*E*' indicates that the groups are on opposite sides of the double bond, and '*Z*' shows that they are on the same side.

Now test yourself

TESTED

1 The alkene pent-2-ene shows *E–Z* (geometrical) isomerism. Draw the displayed formulae of the two stereoisomeric forms.

Answer on p. 219

The Cahn-Ingold-Prelog rules for assigning *E–Z* stereoisomers

Some molecules containing carbon–carbon double bonds have substituents that are all different. How do we decide whether they are *E* or *Z* isomers in these cases?

Method

1. Assign an atomic number to the atom that is directly attached to the carbon atom of the double bond
2. Note down which two atoms have the higher atomic number — these are the important atoms that are given a priority. Sometimes, the atoms may be the same — for example with $-CH_2OH$ and $-CH_2CH_2OH$. In this case, the next bonded atom counts; here oxygen ($Z = 8$) has a higher priority than carbon ($Z = 6$).
3. Which side of the double bond are the priority atoms? If they are on the same side, then it is a *Z* isomer. If they are on opposite sides, then it is an *E* isomer.

Look at the molecules in Figure 12.3. The two priority groups are those of the higher atomic number — chlorine ($Z = 17$) and bromine ($Z = 35$).

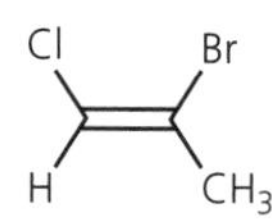

Z-2-bromochloropropene

Cl CH3 H Br

E-2-bromochloropropene

Figure 12.3

> **Exam tip**
>
> Atomic numbers are used to assign priorities to each element, not mass numbers

The isomer on the left is *Z* (or *cis*) as the bromine and chlorine atoms are on the same side of the double bond. However, the molecule on the right has these two atoms on opposite sides, so it is an *E* (or *trans*) isomer.

Now test yourself

TESTED

2 Deduce whether the molecules in Figure 12.4 are *E* or *Z* geometrical isomers using the Cahn-Ingold-Prelog priority rules.

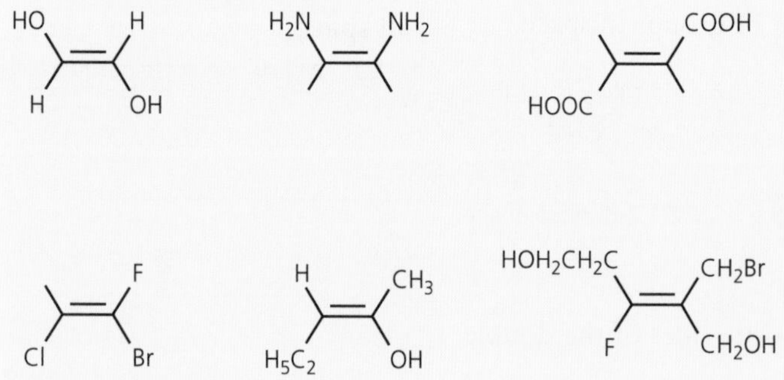

HO H H OH

H_2N NH_2

COOH HOOC

F Cl Br

H CH_3 H_5C_2 OH

HOH_2CH_2C CH_2Br F CH_2OH

Figure 12.4

Answer on p. 219

Addition reactions of alkenes

Electrophilic addition

REVISED

Alkenes react with other substances in a process known as **addition**.

In these reactions, the electron pair present in the π bond of the double bond is donated to the attacking species to form a new single σ bond. The attacking species is therefore an electron pair acceptor, or **electrophile**.

An **electrophile** is an acceptor of a lone pair of electrons.

The mechanism that describes reactions of this type is called **electrophilic addition**.

Bromine, Br_2, reacts with an alkene in this way. For example, ethene, C_2H_4, reacts with bromine to form 1,2-dibromoethane:

$$H_2C{=}CH_2 + Br{-}Br \rightarrow \underset{\text{1,2-dibromoethane}}{H_2CBr{-}CH_2Br}$$

The mechanism by which this reaction takes place is shown in Figure 12.5.

The dotted line is a construction line between the two atoms forming the new bond

Figure 12.5 An electrophilic addition reaction

The reaction of bromine with an alkene is used as a chemical test for unsaturation. In this test, the colour changes from *orange* to *colourless*.

Exam tip

The use of bromine as a test for a carbon–carbon double bond is an important chemical test you need to know.

Reactions of alkenes

REVISED

Alkenes undergo many reactions — for example, the reaction with bromine just mentioned — the other key ones are as follows.

Hydrogen bromide, HBr

Hydrogen bromide *adds* to the double bond in the same way that bromine does, via an **electrophilic addition** process. For example, ethene, C_2H_4, reacts with hydrogen bromide to form bromoethane, C_2H_5Br:

$$H_2C{=}CH_2 + H{-}Br \rightarrow H_3C{-}CH_2Br$$

In this mechanism (Figure 12.6), the first step involves protonation of the carbon–carbon double bond to form a **carbocation**. In the second step, the bromide ion attacks the carbocation.

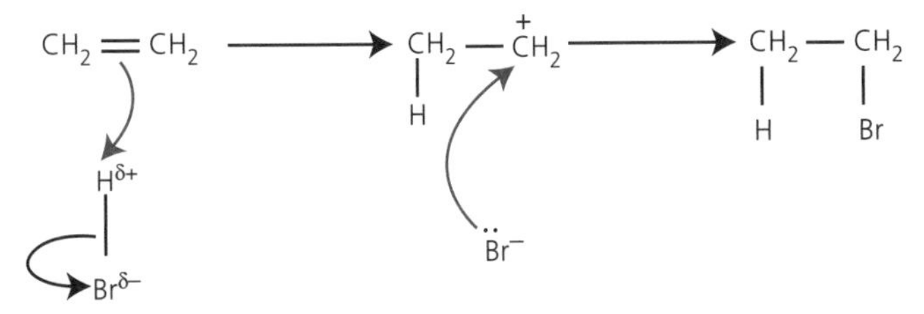

Figure 12.6 Electrophilic addition of HBr to ethene

If hydrogen bromide is added to an unsymmetrical alkene, such as propene, two products can form: 1-bromopropane or 2-bromopropane (Figure 12.7). This is because either carbon atom can be protonated in the first step of the mechanism, leaving the bromide ion free to bond to either of the resulting **carbocations**.

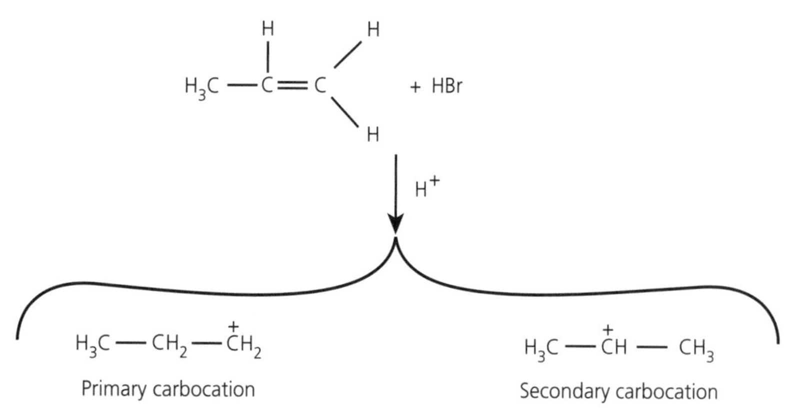

Figure 12.7 Electrophilic addition of HBr to propene

However, primary carbocations are less stable than secondary carbocations and so the latter are formed preferentially. The bromide ion therefore reacts to form 2-bromopropane as the major product.

Example

Draw the product expected when but-1-ene reacts with hydrogen bromide.

Answer

But-1-ene is shown below. The carbon atom coloured red is attached to fewer hydrogen atoms than the other carbon atom in the double bond:

$H_2C{=}CH{-}CH_2CH_3$

In the product, this atom will be bonded to the bromine and the hydrogen will be bonded to the other carbon from the double bond. $H_3C{-}CHBr{-}CH_2CH_3$, 2-bromobutane, is the major product. This is formed via the more stable secondary carbocation (Figure 12.8).

```
    H        H   H
    |   +    |   |
H — C — C — C — C — H
    |   |    |   |
    H   H    H   H
```

Figure 12.8

Now test yourself

TESTED

3 (a) Draw the structure of 2-methylpropene.
(b) Draw the structures of the two carbocations that could form when 2-methylpropene reacts with hydrogen bromide.
(c) Which of the carbocations in (b) will be the more stable?
(d) Hence draw the structure of the major product expected in the reaction.

Answer on p. 219

Sulfuric acid

Alkenes react with concentrated sulfuric acid at 0°C in an electrophilic addition process. The products are hydrogensulfates:

$C_2H_4 + H_2SO_4 \rightarrow H_3C\text{-}CH_2\text{-}OSO_3H$
Ethyl hydrogensulfate

The product is formed via a carbocation intermediate (Figure 12.9), the same as in the reaction with hydrogen bromide.

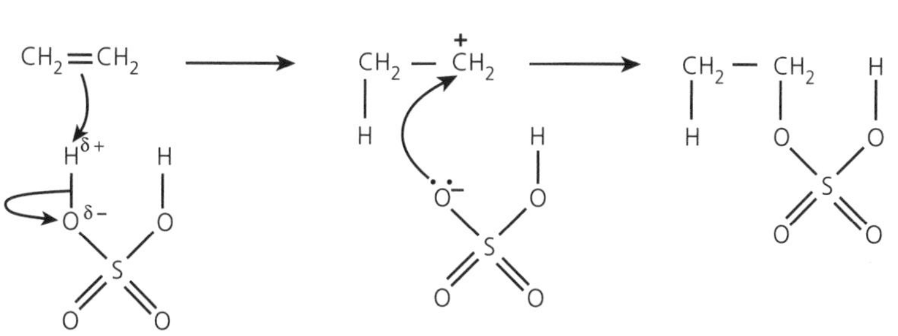

Figure 12.9 Electrophilic addition of H_2SO_4 to ethene

The product reacts with water — a process called hydrolysis — to form an **alcohol**. For example, ethyl hydrogensulfate forms ethanol:

$H_3C\text{-}CH_2\text{-}OSO_3H + H_2O \rightarrow C_2H_5OH + H_2SO_4$

Alcohols are produced industrially by the reaction of steam with an alkene in the presence of an acid catalyst. This process is called **hydration**.

The conditions used are 300°C and 60–70 atm, with phosphoric(v) acid as a catalyst. For example, ethanol can be made by the hydration of ethene using steam:

$C_2H_4(g) + H_2O(g) \rightarrow C_2H_5OH(g)$

Revision activity

There are several alkene reactions you need to know. Test your understanding by drawing the structure of any alkene, and then drawing the products that this molecule would form with HBr, Br_2, H_2SO_4 and then water.

Polymerisation

Addition polymers

REVISED

Alkenes will **polymerise** under high pressure and high temperature conditions and using a range of catalysts. In this process the individual alkene molecules, called **monomers**, bond together (using their π bonds) to form an **addition polymer**. For example, ethene, C_2H_4, forms polyethene in which the two electrons in the π bonds of the monomer (ethene) form σ (single) bonds with other ethene units, as shown in Figure 12.10.

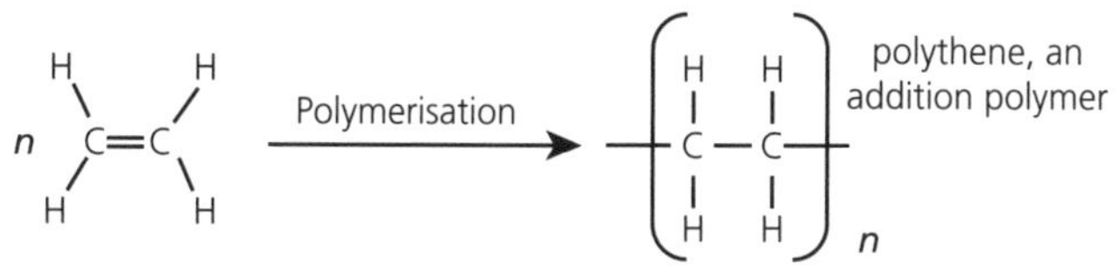

Figure 12.10 Polymerisation

Exam tip

In most polymers, there are no carbon–carbon double bonds. Therefore polymers are fairly unreactive substances.

Figure 12.11 shows how chloroethene (monomer) forms a polymer called polyvinylchloride (PVC).

```
Cl    H          Cl    H          Cl    H
  \  /             \  /             \  /
   C=C       +      C=C       +      C=C
  /  \             /  \             /  \
H     H          H     H          H     H

                   |
                   v

      Cl  H   Cl  H   Cl  H
      |   |   |   |   |   |
 ... —C— —C— —C— —C— —C— —C— ...
      |   |   |   |   |   |
      H   H   H   H   H   H
```

Figure 12.11 Making PVC

Polymers have many uses — film wrapping, carrier bags, bottles, kitchenware, shrink-wrap etc. However, disposal is a problem because many polymers do not biodegrade. Recycling poly(propene), for example, is a more cost-effective method than producing new polymer from crude oil. This is because energy costs are considerably less.

Now test yourself

TESTED

4 Draw the structures of the major products expected when propene reacts with:
 (a) hydrogen bromide, HBr
 (b) bromine, Br_2
 (c) sulfuric acid, H_2SO_4
 (d) steam and an acid catalyst
5 Draw the structure of the polymer polypropene, showing three repeat units.

Answers on p. 219

Exam practice

1 (a) State what is observed when bromine water is added to (i) an alkene and (ii) an alkane. [2]
 (b) Name the mechanism for the reaction between an alkene and a halogen such as bromine. [1]
 (c) Draw the mechanism to show the reaction between propene and bromine. Take care when indicating the positions of any arrows in the mechanism. [4]
2 (a) Name the alkenes in Figure 12.12. [3]

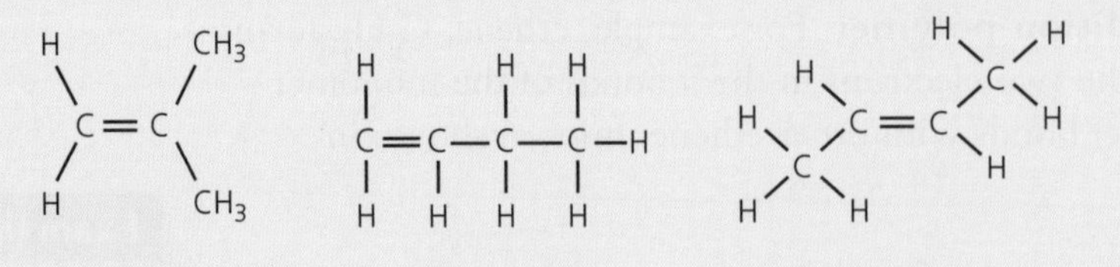

Figure 12.12

 (b) Give the common molecular formula for each of these alkenes. [1]
 (c) How are all three molecules related? [1]
 (d) Explain which of the molecules is able to exist as *E–Z* isomers. [2]
 (e) Draw the structure of the major product formed when molecule 2 reacts with hydrogen bromide. Explain, using the mechanism of the reaction, why this is the major product. [3]
 (f) Draw structures of the repeat units in the polymers formed by:
  (i) molecule 2
  (ii) molecule 3 [2]

3 A student carried out an experiment to determine the number of C=C double bonds in a molecule of a cooking oil by measuring the volume of bromine water decolorised. The student followed these instructions:

- Use a dropping pipette to add 5 drops of oil to $5.0\,cm^3$ of inert organic solvent in a conical flask.
- Use a funnel to fill a burette with bromine water.
- Add bromine water from a burette to the solution in the conical flask and swirl the flask after each addition to measure the volume of bromine water that is decolorised. The student's results are shown in Table 12.1.

Table 12.1

Experiment	Volume of bromine water/cm^3
1	39.40
2	43.50
3	41.20

(a) In a trial experiment, the student failed to fill the burette correctly so that the gap between the tap and the tip of the burette still contained air. Suggest what effect this would have on the measured volume of bromine water in this trial. Explain your answer. [2]

(b) Other than incorrect use of the burette, suggest a reason for the inconsistency in the student's results. [1]

(c) Outline how the student could improve this practical procedure to determine the number of C=C double bonds in a molecule of the oil so that more consistent results are obtained. [4]

(d) The oil has a density of $0.92\,g\,cm^{-3}$ and each of the 5 drops of oil has a volume of $5.0 \times 10^{-2}\,cm^3$. The approximate M_r of the oil is 885. The concentration of bromine water used was $2.0 \times 10^{-2}\,mol\,dm^{-3}$. Use these data and the results from experiment 1 to deduce the number of C=C double bonds in a molecule of the oil. Show your working. [5]

Answers and quick quiz 12 online

ONLINE

Summary

You should now have an understanding of:

- what is meant by the term 'unsaturated' when applied to alkenes
- the structure of a carbon–carbon double bond and how free rotation is restricted
- the existence of *E*–*Z* isomerism
- the reactions of alkenes with HBr, H_2SO_4 and Br_2
- how two products are possible when HBr reacts with an unsymmetrical alkene, and why one may be preferred over the other
- how an alcohol is produced industrially by the reaction of an alkene with steam in the presence of an acid catalyst
- how alkenes polymerise to form addition polymers

13 Alcohols

Naming alcohols

Number of carbon atoms

REVISED

Alcohols are named according to the number of carbon atoms in their molecules, and also the position of the hydroxyl group, OH.

The molecules in Figure 13.1 are called pentan-1-ol, pentan-2-ol and 2-methylbutan-2-ol respectively.

Alcohols are hydrocarbons that have one or more of their hydrogen atoms replaced by a hydroxyl group (–OH).

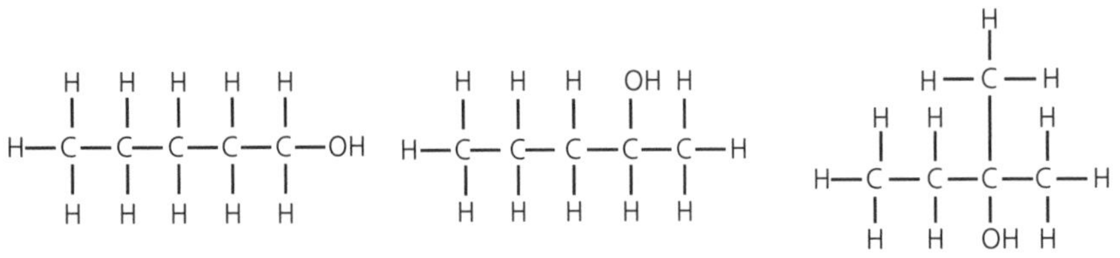

Figure 13.1 Isomeric alcohols

Ethanol production

Ethanol, C_2H_5OH, is a very useful substance because it is a fuel that burns to release heat energy:

$$C_2H_5OH(l) + 3O_2(g) \rightarrow 2CO_2(g) + 3H_2O(l)$$

There are two very different ways of making ethanol: **fermentation of sugars** and **hydration of ethene**.

Fermentation of sugars

REVISED

Sugars are obtained from crops such as sugar cane, sugar beet, corn, rice and maize. Naturally occurring starches (in corn, rice and maize) form sugars by enzyme-controlled processes and, once the sugars are formed, fermentation takes place in the presence of enzymes in yeast:

$$C_6H_{12}O_6(aq) \rightarrow 2C_2H_5OH(aq) + 2CO_2(g)$$

Fermentation is the process in which microorganisms convert one substance into another, normally in the *absence of air*.

Ethanol formed in this way is called a **biofuel** because it is formed from **renewable** plants (those that can be grown again or replenished).

The advantages of this process are as follows:

- Sugars come from **renewable resources**.
- The energy requirements are low (enzymes operate best at room or body temperature).
- The process is carbon-neutral — any carbon dioxide produced is removed from the air by photosynthesis in the original plant so, overall, there has been no net change in carbon dioxide levels.

Disadvantages include the following:

- The ethanol formed is in the form of an aqueous solution — therefore, heat energy is needed for fractional distillation; producing this energy is expensive and not carbon-neutral.

Exam tip

Photosynthesis involves the removal of carbon dioxide from the air; fermentation produces carbon dioxide. So, carbon-neutral fuels have zero effect on carbon dioxide levels.

- The process is slow — it takes several days for the ethanol concentration to reach its limit of approximately 16% by volume.
- Renewable crops require huge surface areas of land to grow, and the machinery and people-power required in harvesting these crops can be significant.

Hydration of ethene

REVISED

This reaction was discussed in Chapter 6 page 54.

Ethene can be hydrated in the presence of phosphoric(v) to form ethanol at 300°C and a pressure of 60–70 kPa (Figure 13.2).

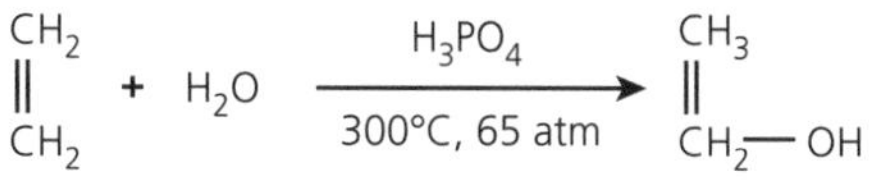

$$CH_2{=}CH_2 + H_2O \xrightarrow[300°C,\ 65\ atm]{H_3PO_4} CH_3{=}CH_2{-}OH$$

Figure 13.2

The advantages of this method are as follows:

- The ethanol is pure, so no separation is required afterwards.
- The rate of reaction is high, so the ethanol is formed quickly.
- The process is continuous, so ethene and steam can be added automatically without any costly breaks in production or use of manpower.

Disadvantages include the following:

- Ethene is obtained from the cracking of crude oil, and cracking requires huge energy reserves.
- Crude oil is a non-renewable resource, so once the crude oil is used it cannot be replenished or reformed.
- The process of hydration requires a high temperature (300°C) and high pressure (60–70 atm) — both of which have considerable energy and cost implications.

Exam tip

Make sure that you know the disadvantages and advantages of the two methods of making ethanol, including the environmental issues.

Classification of alcohols

Alcohols are sorted into groups (Figure 13.3) according to how many alkyl groups are attached to the carbon atom covalently bonded to the hydroxyl group.

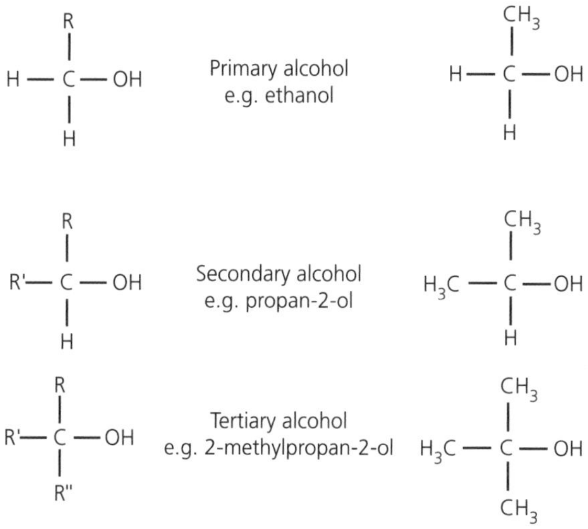

General structure	Class	Example
H—C(R)(H)—OH	Primary alcohol e.g. ethanol	H—C(CH_3)(H)—OH
R'—C(R)(H)—OH	Secondary alcohol e.g. propan-2-ol	H_3C—C(CH_3)(H)—OH
R'—C(R)(R'')—OH	Tertiary alcohol e.g. 2-methylpropan-2-ol	H_3C—C(CH_3)(CH_3)—OH

Figure 13.3 **Primary, secondary and tertiary alcohols**

The class of alcohol determines the type of reaction it may undergo — particularly when being oxidised.

Oxidation of alcohols

REVISED

Alcohols can be oxidised using acidified potassium dichromate(VI) as the oxidising agent. The product molecule formed depends on the class of the original alcohol.

Primary alcohols

Primary alcohols are oxidised to form **aldehydes** and then **carboxylic acids**. For example, ethanol is oxidised in the first stage to form **ethanal**, and then this is oxidised to form **ethanoic acid** — Figures 13.4 and 13.5 show the convenient use of '[O]' to represent the oxidising agent.

$$CH_3CH_2OH + [O] \longrightarrow CH_3CHO + H_2O$$

ethanol → ethanal

$$CH_3CHO + [O] \longrightarrow CH_3COOH$$

ethanal → ethanoic acid

Figure 13.4 Oxidation of primary alcohols

Exam tip

The symbol [O] represents the oxidising agent and is used to balance oxidation equations such as those shown. You should treat it as a normal element symbol, and make sure the oxygens balance.

During these reactions the orange dichromate(VI) ions are **reduced** and turn dark green. The alcohol is **oxidised**. These reactions are therefore **redox reactions**.

- Aldehydes are normally removed by **distillation** (to prevent further oxidation).

Secondary alcohols

As shown in Figure 13.5, secondary alcohols are oxidised to form **ketones**.

$$CH_3CH(OH)CH_3 + [O] \longrightarrow CH_3COCH_3 + H_2O$$

propan-2-ol → propanone

Figure 13.5 Oxidation of secondary alcohols

- Carboxylic acids are formed in a process involving **reflux**. A secondary alcohol and acidified potassium dichromate(VI) are normally **refluxed** to ensure that the reaction goes to completion.

Tertiary alcohols

Tertiary alcohols cannot be oxidised using acidified potassium dichromate(VI) as an oxidising agent. So no reaction occurs and the acidified potassium dichromate(VI) stays orange.

Now test yourself

TESTED

1 Draw the displayed formula of pentan-1-ol.
2 Give the structures and names of the products formed when pentan-1-ol is oxidised using acidified potassium dichromate(VI).

Answers on p. 219

Aldehydes and ketones

Aldehydes and ketones are called **carbonyl compounds** because they contain the carbonyl group. The functional groups present in aldehydes and ketones are shown in Figure 13.6 ('R' and 'R[1]' represent alkyl groups) and some specific molecules in Figure 13.7.

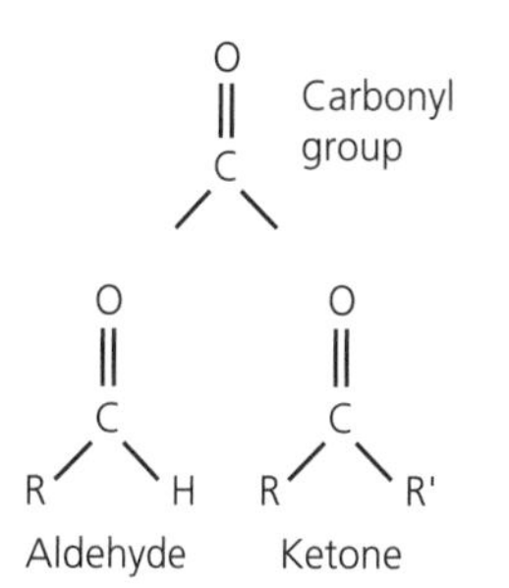

Figure 13.6 Functional groups present in carbonyl compounds

ethanal, CH_3CHO

propanone, CH_3COCH_3

propanal, C_2H_5CHO

pentan-3-one, $CH_3CH_2COCH_2CH_3$

Figure 13.7 Examples of aldehydes and ketones

Distinguishing aldehydes from ketones

REVISED

Tollens' reagent

Aldehydes and ketones can be distinguished from one another using **Tollens' reagent**, which is formed by mixing ammonia with silver nitrate solution.

In this reaction, ketones show no reaction but aldehydes form a **silver mirror** on the inside of the test tube — the silver ions are **reduced**.

$$Ag^+(aq) + e^- \rightarrow Ag(s)$$

and the aldehyde is **oxidised** to a carboxylic acid. Using ethanal as an example:

$$CH_3CHO + [O] \rightarrow CH_3COOH$$

Typical mistake

Many students focus on the reduction of the silver ion when using Tollens' reagent. However, don't forget that the aldehyde has formed a carboxylic acid in the reaction too.

Fehling's solution

In this test, the aldehyde reduces copper(II) ions and an **orange-brown precipitate** of copper(I) oxide is formed. Ketones show no reaction.

Elimination

Dehydration reactions

REVISED

Alcohols can be dehydrated using concentrated sulfuric acid (as a catalyst) to form an alkene. For example, propan-1-ol can be dehydrated to form propene:

$$CH_3CH_2CH_2OH \rightarrow CH_3CH{=}CH_2 + H_2O$$

Dehydration reactions can be used as a possible route for synthesising polymers from alcohols — for example, ethanol → ethene → polyethene.

So polymers could be made starting from sugars, thus avoiding the reliance on crude oil as the original source of carbon.

Now test yourself

TESTED

3 Draw the structures of butanone and butanal.
4 Describe a chemical test to distinguish butanone from butanal.

Answers on p. 219

Exam practice

1 A polymer such as polyethene could be synthesised from glucose using this route:

$$C_6H_{12}O_6\text{ (glucose)} \xrightarrow{\text{step 1}} C_2H_5OH \xrightarrow{\text{step 2}} C_2H_4 \xrightarrow{\text{step 3}} \text{polyethene}$$

(a) Give the reagents and conditions required for each stage of the route above. [6]
(b) Give the name of the chemical process in each step. [3]
(c) Draw the repeat unit in polythene. [1]

2 Substance A in Figure 13.8 can be synthesised from an alcohol under appropriate conditions.

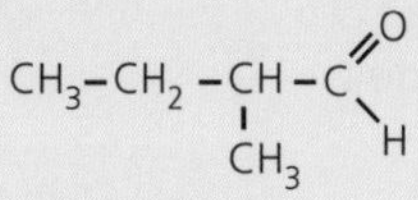

$CH_3–CH_2–CH(CH_3)–CHO$

Figure 13.8

(a) Name molecule A. [1]
(b) Draw the structure of an alcohol that, when oxidised, would produce molecule A. [1]
(c) Give the name of the reagent(s) used to form A from the alcohol. [1]

3 Tollens' reagent is added to a sample of A and the resulting solution heated.
(a) State what would be observed in the reaction. [1]
(b) Draw the structure of the organic molecule formed in the reaction. [1]

4 Compound J, known as leaf alcohol, has the structural formula $CH_3CH_2CH{=}CHCH_2CH_2OH$ and is produced in small quantities by many green plants. The *E*-isomer of J is responsible for the smell of freshly cut grass.
(a) Give the structure of the *E*-isomer of J. [1]
(b) Give the skeletal formula of the organic product formed when J is dehydrated using concentrated sulfuric acid. [1]
(c) Another structural isomer of J is shown in Figure 13.9.

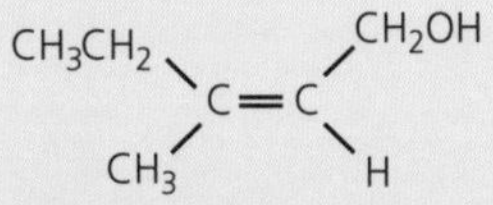

$(CH_3CH_2)(CH_3)C{=}C(CH_2OH)(H)$

Figure 13.9

Explain how the Cahn-Ingold-Prelog (CIP) priority rules can be used to deduce the full IUPAC name of this compound. [6]

Answers and quick quiz 13 online

ONLINE

Summary

You should now have an understanding of:

- the molecular structure of alcohols and their key functional group
- how to name simple alcohols
- how ethanol can be made on an industrial scale, either by fermentation or hydration, and the advantages and disadvantages of each process
- how to classify alcohols as primary, secondary or tertiary
- the oxidation of alcohols of different classes using acidified potassium dichromate(VI)
- Tollens' reagent and how it is used to distinguish between aldehydes and ketones
- elimination reactions, in which alkenes are made from alcohols

14 Organic analysis

Several modern analytical techniques are useful when determining the structures of molecules.

Identification of functional groups by test-tube reactions

You may be asked to carry out simple chemical tests that would enable you to distinguish between well-known functional groups, for example alkenes, alcohols, carboxylic acids, aldehydes and ketones. These chemical tests are already mentioned in earlier chapters, but for convenience, they are summarised in Table 14.1.

Table 14.1

Functional group being tested	Test being carried out	Positive identification
Alkenes $>C=C<$	Bromine water, $Br_2(aq)$, is added to the sample	An orange solution will turn colourless
Alcohols R—O—H	Add acidified potassium dichromate(VI) and warm the mixture gently	(a) Primary and secondary alcohol — an orange solution will turn green (b) Tertiary alcohol — no reaction takes place, so the solution stays orange
Carboxylic acids R—C(=O)—O—H	Test as a typical acid and add some sodium carbonate, Na_2CO_3, either as a solid or solution	Fizzing will occur and the gas can be bubbled through limewater, which goes milky
Aldehydes R—C(=O)—H	Add either (a) ammoniacal silver nitrate solution or (b) Fehling's solution, and warm in a water bath	(a) A silver mirror will form (b) A blue solution will form a brown/orange/red precipitate

Now test yourself

TESTED

1 What are the systematic names of the molecules in Figure 14.1?

CH_3—CH_2—C(=O)—O—H   H—C(H)(H)—C(H)(O—H)—C(H)(H)—H

Figure 14.1

2 Describe a chemical test that would enable you to distinguish between the molecules in Figure 14.1.

3 Draw the structures of, and name, the organic products formed in these two tests.

Answers on p. 219

Mass spectrometry

Mass spectrometry instrumentation was discussed earlier in Chapter 1 where it was used for analysing elements. It is also very useful when analysing organic compounds.

It is possible to measure with great precision the mass of a **molecular ion** if the relative isotopic masses to four decimal places are used. For example:

$^{1}H = 1.0078$ $^{12}C = 12.0000$ $^{14}N = 14.0031$ $^{16}O = 15.9949$

A **molecular ion** is the original molecule that has lost one electron, M^+. Its mass will be virtually the same as the molecule being tested, because an electron has negligible mass.

For example, the two organic compounds propane (C_3H_8) and ethanal (CH_3CHO) both have relative molecular mass $M_r = 44$ to the nearest whole number. Using a high-resolution mass spectrometer, two molecular ion peaks with the following more precise *m/z* values are obtained:

$C_3H_8 = 44.0624$ $CH_3CHO = 44.0261$

When analysing samples, it is therefore possible to deduce the molecular formula of a compound by measuring its precise relative molecular mass. The relative molecular mass of a compound is given by the peak with the highest *m/z* value.

A computer database can hold the relative molecular masses of thousands of organic molecules, measured to a high level of accuracy; a sample under test can then be compared to other spectra and its identity determined.

Infrared spectroscopy

Atoms in covalent bonds are **vibrating** about a mean position. The frequency of the vibration depends on the masses of the atoms in the bond — the greater the mass of the atoms, the slower the vibrations.

Vibrational frequency

REVISED

The frequency of such a vibration can be quoted in **wavenumbers** — the number of waves in 1 cm. One particular bond will vibrate at a certain wavenumber. However, just as the energy of a particular covalent bond will vary slightly in different compounds, the frequency of vibration also varies, so values are usually quoted as a range, as in Table 14.2.

Table 14.2 Typical wavenumbers in infrared spectra

Bond	Wavenumber/cm^{-1}
C–H in alkanes	2850–2960
C–C	750–1100
C–O	1000–1300
O–H in carboxylic acids	2500–3000
C=O	1680–1750
O–H in hydrogen-bonded alcohols	3230–3550 (broad adsorption)
C–N in amines	1180–1360
C–Cl	600–800
C–Br	500–600

Figure 14.2 shows the displayed formula for ethyl ethanoate, $CH_3COOC_2H_5$.

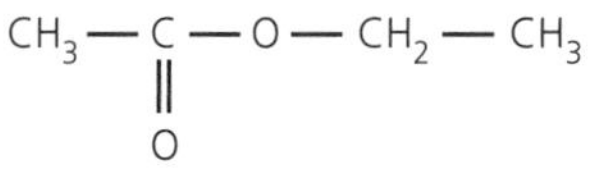

$$CH_3 - \underset{\underset{O}{\|}}{C} - O - CH_2 - CH_3$$

Figure 14.2 **Ethyl ethanoate**

Notice the bonds present in this compound: C–H, C=O, C–O and C–C. These will vibrate at certain wavenumbers. For example, the C=O bond is expected to vibrate at 1743 cm^{-1} (Figure 14.3).

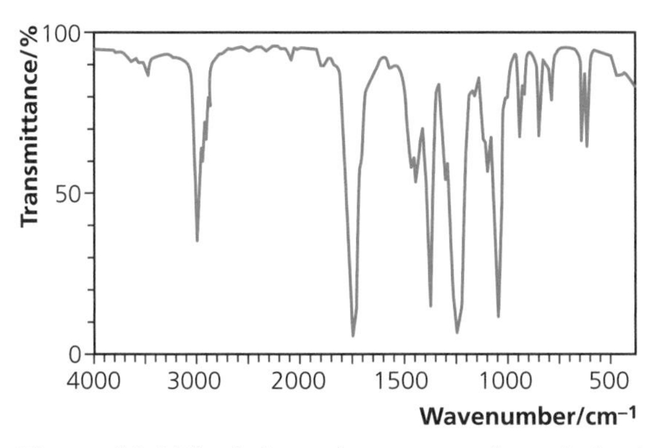

Figure 14.3 **The infrared spectrum for ethyl ethanoate**

The fingerprint region

The peaks below 1500 cm^{-1} in the spectrum are used collectively as a fingerprint region. This characteristic arrangement of peaks is unique to the compound, just as a fingerprint is unique to a person. Therefore, by comparing the fingerprint region of an unknown compound with many on a computer database, the identity of the compound can be deduced.

It is possible to detect impurities in a compound because these may give rise to absorptions that would not be expected for that compound.

Some molecules in the atmosphere absorb infrared radiation and contribute towards global warming. This is because the natural frequency of the bond vibration in molecules such as CO_2 or H_2O coincides exactly with the frequency of the infrared radiation. The same process takes place when a sample is put into an infrared spectrometer.

Now test yourself

TESTED

4 Draw the structure of propan-1-ol.
5 Write down the bonds present in propan-1-ol.
6 Give the wavenumbers of the main peaks expected in the infrared spectrum of this compound.
7 A small absorption is observed at a wavenumber of 1720 cm^{-1}. What does this suggest about the sample?

Answers on p. 219

Exam practice

1 In an experiment to prepare a sample of ethanal (CH_3CHO), ethanol (C_2H_5OH) is reacted with acidified potassium dichromate(VI) and the reaction mixture is distilled. Infrared spectra are obtained for ethanol and ethanal (Figure 14.4).

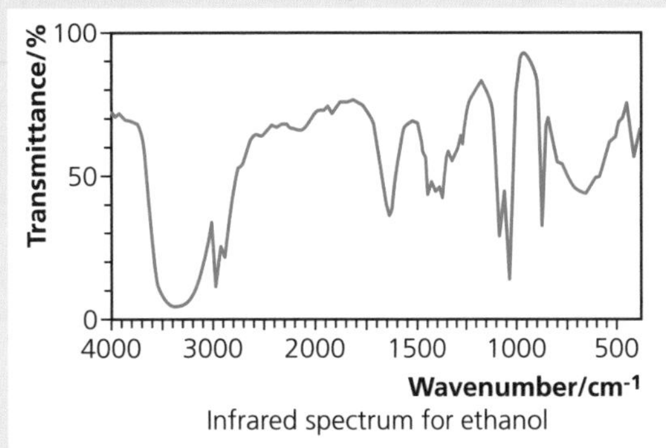

Infrared spectrum for ethanol

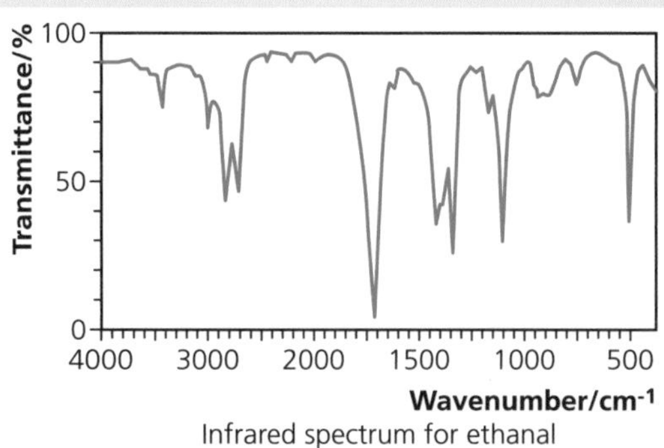

Infrared spectrum for ethanal

Figure 14.4

(a) Indicate the bonds that give rise to these absorptions:
  (i) in the ethanol spectrum at 3400 cm^{-1}
  (ii) in the ethanal spectrum at 1720 cm^{-1} [2]

(b) Write an equation to show the oxidation of ethanol to form ethanal, using [O] to represent the oxidising agent. [1]

(c) Explain why the absorption at 3400 cm^{-1} in the ethanol spectrum does not appear in the spectrum for ethanal. [1]

(d) The reaction mixture from the experiment preparing ethanal was left for several days, and another infrared spectrum taken. It was observed that a new absorption was present between 2500 cm^{-1} and 3000 cm^{-1}. Explain this observation. [2]

2 When cyclohexanol is heated with concentrated sulfuric acid, a reaction occurs and a new product is formed. The infrared spectrum for the new compound is shown in Figure 14.5.

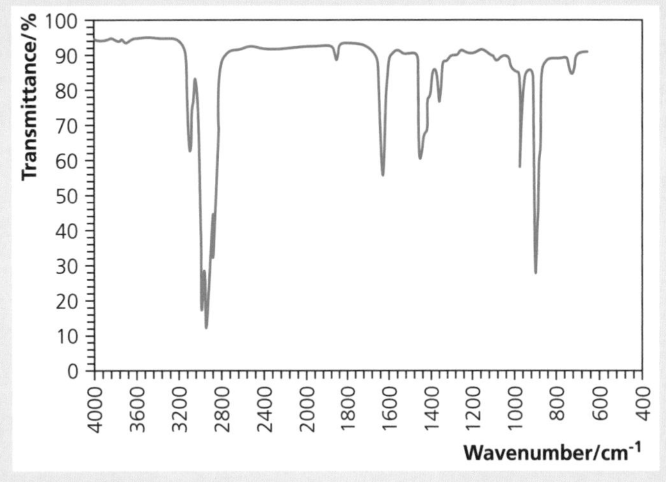

Figure 14.5

(a) Use the data in Table 14.3 to explain how the spectrum can indicate that cyclohexanol is no longer present. [1]

Table 14.3 Infrared absorption data

Bond	Wavenumber/cm^{-1}
C–H	2850–3300
C–C	750–1100
C=C	1620–1680
C=O	1680–1750
C–O	1000–1300
O–H (alcohols)	3230–3550
O–H (acids)	2500–3000

(b) Using the spectrum, identify the product of the reaction, giving reasons for your choice. [2]

(c) Indicate how high-resolution mass spectrometry can be used to identify the product of this reaction. [1]

3 Give two differences between the infrared spectrum of a carboxylic acid and that of an alcohol other than in their fingerprint regions. [2]

Answers and quick quiz 14 online

ONLINE

Summary

You should now have an understanding of:

- how mass spectrometry can be used to identify a substance by measuring an accurate value for the mass of the molecular ion
- how infrared spectroscopy works by measuring characteristic vibrational frequencies in a molecule
- how infrared spectroscopy can be used to identify the main bonds in a molecule
- the 'fingerprint region' in the infrared spectrum and its importance when identifying a compound
- global warming in terms of molecular vibrations — the same process that takes place when measuring the infrared spectrum of a compound

15 Thermodynamics

Enthalpy change

There are some important definitions that you must learn and also apply, some of which you will already know from the AS course. All the energy terms listed below are considered to be **standard values** — that is, their measurement is done at 298 K and a pressure of 100 kPa.

- **Enthalpy of formation** is the enthalpy change when 1 mole of a substance is formed from its constituent elements, with all reactants and product in their standard states. For example, using calcium carbonate:

  $Ca(s) + C(s) + 1\frac{1}{2}O_2(g) \rightarrow CaCO_3(s)$

- **First ionisation enthalpy** is the enthalpy change for the removal of 1 mole of electrons from 1 mole of gaseous atoms to produce 1 mole of singly charged positive ions in the gas phase. For example, using magnesium:

  $Mg(g) \rightarrow Mg^+(g) + e^-$

- **Enthalpy of atomisation** is the enthalpy change for the formation of 1 mole of gaseous atoms from an element in its standard state. For example, using oxygen:

  $\frac{1}{2}O_2(g) \rightarrow O(g)$

- **Bond dissociation enthalpy** is the enthalpy change required to break 1 mole of a specific bond in a specific compound in the gas phase. For example, using ammonia:

  $NH_3(g) \rightarrow NH_2(g) + H(g)$

- **Mean bond dissociation enthalpy** is the average of the bond dissociation energies for a given bond measured over many compounds.
- **Electron affinity** is the enthalpy change when 1 mole of electrons is added to 1 mole of gaseous atoms to form 1 mole of gaseous negative ions. For example, using chlorine:

  $Cl(g) + e^- \rightarrow Cl^-(g)$

- **Lattice formation enthalpy** is the enthalpy change when 1 mole of an ionic solid is formed from its constituent gaseous ions. For example, using magnesium chloride:

  $Mg^{2+}(g) + 2Cl^-(g) \rightarrow MgCl_2(s)$

- **Lattice dissociation enthalpy** is the enthalpy change when 1 mole of an ionic solid is turned into its constituent gaseous ions. For example, using magnesium chloride:

  $MgCl_2(s) \rightarrow Mg^{2+}(g) + 2Cl^-(g)$

- **Enthalpy of hydration** is the enthalpy change when 1 mole of a gaseous species dissolves in an infinite volume of water. For example, using lithium ions:

  $Li^+(g) + aq \rightarrow Li^+(aq)$

Exam tip

Learn these definitions — they are often requested in examinations — but also understand their chemical significance. There is often a change of physical state and this will make a huge difference, so do take note of any state symbols.

Typical mistake

Many students realise that most of the definitions on this page are defined according to *starting* with 1 mole of substance. However, enthalpy of atomisation is defined according to the *formation* of 1 mole of gaseous atoms. Be careful.

- **Enthalpy of solution** is the enthalpy change when 1 mole of a substance is dissolved in water to infinite dilution. For example, using sodium sulfate(VI):

  $Na_2SO_4(s) \rightarrow 2Na^+(aq) + SO_4^{2-}(aq)$

The Born–Haber cycle

REVISED

A lattice enthalpy can be analysed or calculated using a **Born–Haber cycle** like that shown in Figure 15.1.

Example

Draw a Born–Haber cycle for the formation of calcium fluoride, CaF_2, and use it to calculate its lattice formation enthalpy.

Answer

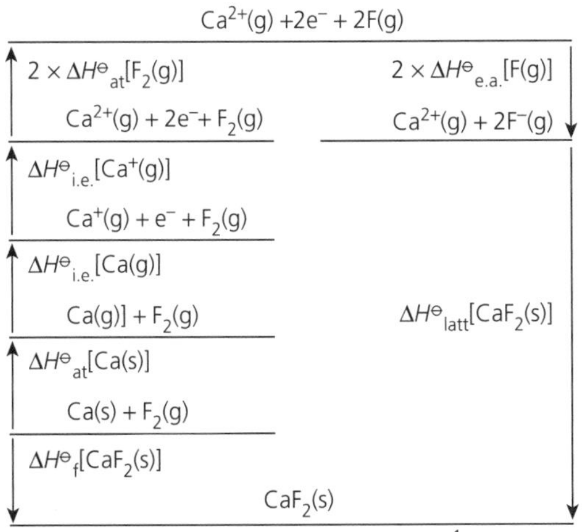

$Ca^{2+}(g) + 2e^- + 2F(g)$

$2 \times \Delta H^\ominus_{at}[F_2(g)]$ $\qquad$ $2 \times \Delta H^\ominus_{e.a.}[F(g)]$

$Ca^{2+}(g) + 2e^- + F_2(g)$ $\qquad$ $Ca^{2+}(g) + 2F^-(g)$

$\Delta H^\ominus_{i.e.}[Ca^+(g)]$

$Ca^+(g) + e^- + F_2(g)$

$\Delta H^\ominus_{i.e.}[Ca(g)]$

$Ca(g)] + F_2(g)$ $\qquad$ $\Delta H^\ominus_{latt}[CaF_2(s)]$

$\Delta H^\ominus_{at}[Ca(s)]$

$Ca(s) + F_2(g)$

$\Delta H^\ominus_f[CaF_2(s)]$

$CaF_2(s)$

Relevant standard enthalpy data (in $kJ\,mol^{-1}$):

$\Delta H^\ominus_f[CaF_2(s)] = -1214$

$\Delta H^\ominus_{at}[Ca(s)] = +193$

$\Delta H^\ominus_{i.e.}[Ca(g)] = +590$

$\Delta H^\ominus_{i.e.}[Ca^+(g)] = +1150$

$\Delta H^\ominus_{at}[F_2(g)] = +79$

$\Delta H^\ominus_{e.a}[F(g)] = -348$

Figure 15.1 Born–Haber cycle for CaF_2

Using the cycle:

$$\Delta_{at}H^\ominus[Ca(s)] + \Delta_{i.e.}H^\ominus[Ca(g)] + \Delta_{i.e.}H^\ominus[Ca^+(g)] + (2 \times \Delta_{at}H^\ominus[F_2(g)]) + (2 \times \Delta_{e.a.}H^\ominus[F(g)]) + \Delta_{latt}H^\ominus[CaF_2(s)] = \Delta_f H^\ominus[CaF_2(s)]$$

$$\Delta_{latt}H^\ominus[CaF_2(s)] = \Delta_f H^\ominus[CaF_2(s)] - \Delta_{at}H^\ominus[Ca(s)] - \Delta_{i.e.}H^\ominus[Ca(g)] - \Delta_{i.e.}H^\ominus[Ca^+(g)] - (2 \times \Delta_{at}H^\ominus[F_2(g)]) - (2 \times \Delta_{e.a.}H^\ominus[F(g)])$$

$$= (-1214) - (+193) - (+590) - (+1150) - (2 \times +79) - (2 \times -348)$$

$$= -1214 - 193 - 590 - 1150 - 158 + 696$$

$$= -2609\,kJ\,mol^{-1}$$

Notice that the lattice energy is **highly exothermic**. This is because attractions are being formed between the gaseous ions as they come together to form their ionic lattice.

It is likely that questions will be set in examinations in which you are asked to calculate any one of the terms in a Born–Haber cycle given all others.

Exam tip

When calculating an enthalpy change using a cycle such as the Born–Haber cycle, apply Hess's law to find an alternative route. Remember to change the sign of any term that points in the opposite direction from the one you want to move along.

Uses of lattice enthalpy

Lattice enthalpy is related to the **charges** on the ions and their **ionic radii**. The higher the ionic charges and the smaller the ionic radii, the more exothermic will be the lattice formation enthalpy. For example, sodium fluoride has a value for $\Delta_{latt}H^{\ominus}$ of $-902\,kJ\,mol^{-1}$, whereas magnesium oxide has a value of $-3889\,kJ\,mol^{-1}$. The Mg^{2+} ion is smaller than the Na^+ ion, and it also has a higher charge; the O^{2-} ion is a similar size to the F^- ion but it has a higher negative charge. Therefore, the Mg^{2+} and O^{2-} ions will be more strongly attracted in MgO than the Na^+ and F^- ions in NaF. MgO therefore has a more exothermic lattice formation enthalpy.

It is possible to calculate a **theoretical value** for the lattice formation enthalpy of an ionic solid — this assumes that the solid is 100% ionic in character. When such values are compared with those obtained using a Born–Haber cycle (for which the values are obtained experimentally), evidence can be gained about the type of bonding present in a compound.

If there is a significant difference between the 100% ionic model and the experimental Born–Haber cycle value (Table 15.1) then we can assume that there must be some **covalent character** in the substance. This is attributed to the **polarising power** of small and highly charged positive ions distorting the spherical electron distribution of the negative ions, thereby **inducing** some covalent character and strengthening the lattice.

Table 15.1 Covalent character in ionic compounds

Substance	Theoretical lattice enthalpy/$kJ\,mol^{-1}$	Born–Haber lattice enthalpy/$kJ\,mol^{-1}$	Difference/%
LiCl	833	846	1.56
NaCl	766	771	0.65
KCl	690	701	1.59
RbCl	674	675	0.15
AgCl	770	905	17.5
AgBr	758	890	18.7
AgI	736	876	19.0

The following deductions can be made:

- The ionic model fits better at the bottom of Group 1 than at the top. Lithium ions are smaller than the other ions in the group, and so they will have a bigger polarising power towards the chloride ions, hence inducing more covalent character in the compound.
- There is more **ionic character** in the compounds as Group 1 is descended.
- There is, conversely, more covalent character moving up the group.
- Differences are much more marked in other regions of the periodic table.

Lattice enthalpies can also be used for calculating the **enthalpy of solution** of a compound (Figure 15.2). An energy cycle is used that combines the lattice enthalpy and the hydration enthalpies of the ions with the enthalpy of solution:

$Na^+(g) + Cl^-(g)$

$\Delta H_{latt}[NaCl(s)]$

$NaCl(s)$

$\Delta H_{sol}[NaCl(s)]$

$\Delta H_{hyd}[Na^+(g)] + \Delta H_{hyd}[Cl^-(g)]$

$Na^+(aq) + Cl^-(aq)$

Figure 15.2 Calculating enthalpy of solution

$\Delta_{latt}H^{\ominus}[NaCl(s)] = -771\ kJ\ mol^{-1}$; $\Delta_{hyd}H^{\ominus}[Na^+(g)] = -406\ kJ\ mol^{-1}$; $\Delta_{hyd}H^{\ominus}[Cl^-(g)] = -364\ kJ\ mol^{-1}$

You can see from the cycle that:

$$\Delta_{sol}H^{\ominus}[NaCl(s)] = \Delta_{hyd}H^{\ominus}[Na^+(g)] + \Delta_{hyd}H^{\ominus}[Cl^-(g)] - \Delta_{latt}H^{\ominus}[NaCl(s)]$$

If the sum of the hydration enthalpies is greater than the lattice enthalpy, then the substance will have a negative value for the enthalpy of solution, and vice versa.

For sodium chloride:

$$\Delta_{sol}H^{\ominus}[NaCl(s)] = (-406) + (-364) - (-771)$$

$$= -406 - 364 + 771$$

$$= +1\ kJ\ mol^{-1}$$

The enthalpy of solution for sodium chloride is therefore only slightly endothermic — this is because the lattice enthalpy is balanced by the sum of the hydration enthalpies almost exactly.

Exam tip

Lattice enthalpies and hydration enthalpies depend on the same factors — the higher the ionic charges and the smaller the ionic radii, the greater will be the attraction between the ions in the lattice, and the attraction between water molecules and the individual ions.

Now test yourself

TESTED

The ionic radii, in nm, of some positive and negative ions are given in the table. Use the data to answer the questions that follow.

Positive ions	Ionic radii/nm	Negative ions	Ionic radii/nm
K^+	0.133	F^-	0.136
Ca^{2+}	0.100	Cl^-	0.188
		Br^-	0.195
		S^{2-}	0.185

1 Predict which of the following compounds will have the *largest* difference between its lattice energy calculated from experimental values and the theoretical value calculated assuming 100% ionic character.
A CaS B K_2O C KF D CaO

2 Predict which of the following will have the most exothermic lattice formation enthalpy. Assume that they all have the same crystal structure.
A CaS B CaO C KF D KCl

3 Predict which of the ions below is likely to have the most exothermic hydration enthalpy.
A F^- B Ca^{2+} C Br^- D K^+

Answers on p. 220

Bond enthalpies

REVISED

Bond dissociation enthalpy (b.e.) is the enthalpy change required to break 1 mole of a specific bond in a specific compound in the gas phase.

There is often a big difference between the bond strength in a particular compound and the average value worked out over many compounds.

Nearly always in calculations, except for elements, it is the mean values that are quoted (in $kJ\,mol^{-1}$). For example:

N–H = 388 H–Cl = 431 O–H = 463 N–N = 145 H–H = 436

O–O = 157 O=O = 496 Cl–Cl = 242 N≡N = 944

Example 1

Calculate the enthalpy change for the reaction that takes place between hydrazine and hydrogen peroxide:

$$N_2H_4(g) + 2H_2O_2(g) \rightarrow N_2(g) + 4H_2O(g)$$

In terms of bonds, this looks like:

H₂N–NH₂ + 2 H–O–O–H ⟶ N≡N + 4 H–O–H

Answer

Using Hess's law, we can draw this energy cycle:

$N_2H_4(g) + 2H_2O_2(g) \xrightarrow{\Delta H_1} N_2(g) + 4H_2O(g)$

ΔH_2: $N_2H_4(g) + 2H_2O_2(g) \rightarrow 2N(g) + 8H(g) + 4O(g)$

ΔH_3: $N_2(g) + 4H_2O(g) \rightarrow 2N(g) + 8H(g) + 4O(g)$

It can be seen that $\Delta H_1 = \Delta H_2 - \Delta H_3$.

Using the bond energies above:

ΔH_2 = b.e.(N–N) + 4 × b.e.(N–H) + 2 × b.e.(O–O) + 4 × b.e.(O–H)

ΔH_3 = b.e.(N≡N) + 8 × b.e.(O–H)

So:

ΔH_1 = (145) + (4 × 388) + (2 × 157) + (4 × 463) – (944) – (8 × 463)

= 145 + 1552 + 314 + 1852 – 944 – 3704

= $-785\,kJ\,mol^{-1}$

Example 2

Calculate the enthalpy change for the reaction below in which substances are all in their standard states:

$N_2H_4(l) + 2H_2O_2(l) \rightarrow N_2(g) + 4H_2O(l)$

Enthalpy of formation data:

$\Delta_f H^\ominus$/kJ mol^{-1}: $N_2H_4(l) = +50$; $H_2O_2(l) = -188$; $H_2O(l) = -286$

Answer

From the AS course:

$$\Delta H = \Sigma\Delta_f H_{products} - \Sigma\Delta_f H_{reactants}$$

$$= [(4 \times -286) + 0] - [(+50) + (2 \times -188)]$$

$$= -1144 + 326$$

$$= -818\,\text{kJ mol}^{-1}$$

Comparing this value with that obtained in example 1, it can be seen that there is a difference of 33 kJ mol^{-1}. The reasons for this include:

- In example 1, mean bond enthalpies were used for compound bonds when specific bond enthalpies for the substances in the reaction would have been more accurate.
- In example 1, hydrogen peroxide, hydrazine and water were all in the gas phase. Their standard state is liquid.

Now test yourself

TESTED

4 Calculate ΔH for the reactions below using bond enthalpies given in this section.
   (a) $N_2(g) + 2H_2(g) \rightarrow N_2H_4(g)$
   (b) $H_2O_2(g) \rightarrow H_2(g) + O_2(g)$
   (c) $H_2O_2(g) \rightarrow H_2O(g) + \frac{1}{2}O_2(g)$

5 The enthalpy of formation of hydrazine, from:
   $N_2(g) + 2H_2(g) \rightarrow N_2H_4(l)$
   is +50.4 kJ mol^{-1}.
   Comment on any differences between this value and that from question 4(a).

Answers on p. 220

Entropy and Gibbs free energy

Entropy, S, is a measure of the amount of **disorder** of the particles being considered. For example, the standard entropies ($S^\ominus$), measured at 298 K, of three elements are shown in Table 15.2.

The **entropy** of the three main physical states increases in the order solid to liquid to gas.

Table 15.2 Entropy values for some elements

Element	$S^\ominus$/J K^{-1} mol^{-1}
Fe(s)	27.2
Hg(l)	77.4
O_2(g)	205

It can be seen that the values increase on moving from a solid to a liquid to a gas. This is because there is more disorder in a gaseous system and there will, therefore, be a greater number of ways in which the energy possessed by the molecules present can be arranged.

Many processes have associated with them a spontaneous **increase of disorder**, or increase in entropy — for example:

- **Melting**

  $H_2O(s) \rightarrow H_2O(l)$
- **Evaporation**

  $C_2H_5OH(l) \rightarrow C_2H_5OH(g)$
- **Dissolution** or dissolving

  $CaCl_2(s) + (aq) \rightarrow Ca^{2+}(aq) + 2Cl^-(aq)$
- **Formation of a gas**

  $NaHCO_3(s) + HCl(aq) \rightarrow NaCl(aq) + H_2O(l) + CO_2(g)$

Now test yourself

TESTED

6 State whether each of the following processes involves an increase or decrease in entropy.

(a) $CuSO_4(s) + 5H_2O(l) \rightarrow CuSO_4.5H_2O(s)$

(b) $Mg(s) + H_2SO_4(aq) \rightarrow MgSO_4(aq) + H_2(g)$

(c) $NaCl(s) + (aq) \rightarrow NaCl(aq)$

(d) $2CO(g) + O_2(g) \rightarrow 2CO_2(g)$

(e) $H_2O(l) \rightarrow H_2O(s)$

Answers on p. 220

Calculating entropy changes

When a process happens, it is possible to calculate the entropy change by working out the difference between the final entropy and the initial entropy. For example, under standard conditions:

$$\Delta S^\ominus = \Sigma S^\ominus_{products} - \Sigma S^\ominus_{reactants}$$

Example

Calculate the standard entropy change for the following process, given the individual absolute standard entropy values:

$$2NaHCO_3(s) \rightarrow Na_2CO_3(s) + H_2O(l) + CO_2(g)$$

$S^\ominus$/J K^{-1} mol^{-1}	102	136	70	214

$$\Delta S^\ominus = \Sigma S^\ominus_{products} - \Sigma S^\ominus_{reactants}$$

$$\Delta S^\ominus = (136 + 70 + 214) - (2 \times 102)$$

$$= 420 - 204$$

$$= +216\ J\ K^{-1}\ mol^{-1}$$

The entropy change is highly positive — this was predictable because a gas was formed. An increase in the number of moles of gas always leads to an increase in entropy.

Typical mistake

In questions like this, many students forget to multiply the $NaHCO_3$ entropy by 2.

Now test yourself

TESTED

7 (a) Calculate the standard entropy change, $\Delta S^\ominus$, for the reaction:

	$CaO(s)$ +	$CO_2(g)$ →	$CaCO_3(s)$
$S^\ominus/J\,K^{-1}\,mol^{-1}$	40	214	93

(b) Comment on the sign of the entropy value and explain why this should be expected.

Answers on p. 220

Gibbs free energy, ΔG

REVISED

Enthalpy change alone cannot be used to predict whether or not a reaction will take place on its own. Reactions that are spontaneous are also described as **feasible**.

Although many reactions that take place readily are exothermic, there are also examples of endothermic reactions that are spontaneous — for example, the dissolving of barium nitrate, $Ba(NO_3)_2(s)$, in water:

$$Ba(NO_3)_2(s) + (aq) \rightarrow Ba^{2+}(aq) + 2NO_3^-(aq) \qquad \Delta H = +40.4\,kJ\,mol^{-1}$$

Exam tip

ΔG must be negative for a reaction to occur spontaneously.

Clearly, a better way of determining whether reactions are spontaneous or not is required. This uses the **Gibbs free energy change**, ΔG, which is the true indicator of the expected direction of chemical change.

A Gibbs free energy change is calculated using the relationship:

$$\Delta G = \Delta H - T\Delta S$$

- ΔG is the Gibb's free energy change in $kJ\,mol^{-1}$.
- ΔH is the molar enthalpy change for the reaction in $kJ\,mol^{-1}$.
- T is the Kelvin temperature.
- ΔS is the entropy change in $J\,K^{-1}\,mol^{-1}$.

Example

Given the data provided, calculate $\Delta G^\ominus$, at 298 K, for the reaction:

	$ZnO(s)$ +	$CO(g)$ →	$Zn(s)$ +	$CO_2(g)$
$S^\ominus/J\,K^{-1}\,mol^{-1}$	44	198	42	214
$\Delta H^\ominus_f/kJ\,mol^{-1}$	−348	−111	0	−394

Answer

Calculate the standard entropy change for the reaction using $\Delta S^\ominus = \Sigma S^\ominus_{products} - \Sigma S^\ominus_{reactants}$

$$\Delta S^\ominus = (42 + 214) - (44 + 198)$$
$$= 256 - 242$$
$$= +14\,J\,K^{-1}\,mol^{-1}$$

Calculate the enthalpy change for the reaction using $\Delta H^\ominus = \Sigma H^\ominus_{products} - \Sigma H^\ominus_{reactants}$

$$\Delta H^\ominus = (-394 + 0) - (-348 + -111)$$
$$= -394 - (-459)$$
$$= -394 + 459$$
$$= +65\,kJ\,mol^{-1}$$

Calculate the standard Gibbs free energy change using $\Delta G^\ominus = \Delta H^\ominus - T\Delta S^\ominus$:

$$\Delta G^\ominus = +65 - \frac{298 \times 14}{1000}$$

$$= 65 - 4.2$$

$$= +60.8\,\text{kJ}\,\text{mol}^{-1}$$

This reaction is not spontaneous at 298 K because the sign for $\Delta G^\ominus$ is positive. So, it is not possible for carbon monoxide to reduce zinc oxide to form zinc and carbon dioxide at 298 K.

Typical mistakes

Many students forget to divide the entropy term by 1000 to convert from J to kJ when they use $\Delta G = \Delta H - T\Delta S$.

Deducing the temperature at which a reaction becomes spontaneous

Another term for 'spontaneous' is 'feasible'.

Looking at the equation $\Delta G = \Delta H - T\Delta S$, it can be seen that:

- as the temperature increases, $T\Delta S$ becomes more positive
- as $T\Delta S$ becomes more positive, it must reach a temperature at which $T\Delta S$ exceeds ΔH
- at this temperature, ΔG will start to be negative and the reaction is now spontaneous.

Example 1

Using this reaction again:

$$ZnO(s) + CO(g) \rightarrow Zn(s) + CO_2(g)$$

At what temperature does it become feasible?

Answer

At the temperature at which ΔG changes from being positive to negative, $\Delta G = 0$ and:

$$\Delta H = T\Delta S$$

So:

$$T = \frac{\Delta H}{\Delta S}$$

But:

$$\Delta H = +65\,\text{kJ}\,\text{mol}^{-1} \text{ and } \Delta S = +0.014\,\text{kJ}\,\text{K}^{-1}\,\text{mol}^{-1}$$

$$T = \frac{65}{0.014}$$

$$= 4643\,\text{K or } 4370°\text{C}$$

This means that carbon monoxide can reduce zinc oxide at or above 4370°C. Such a high temperature is very expensive to produce, so the process is probably an uneconomic method for producing zinc.

Example 2

(a) Calculate ΔG for the following reaction at 298 K given the data provided below.

$4NH_3(g) + 5O_2(g) \rightarrow 4NO(g) + 6H_2O(g)$

Substance	NH_3	O_2	NO	H_2O
$S°$/J K^{-1} mol^{-1}	193	205	211	189
$\Delta H°$/kJ mol^{-1}	-46.2	0	90.4	-242

(b) State whether the reaction will be spontaneous or not at 298 K.

Answer

(a) $\Delta G° = \Delta H° = T\Delta S°$

$\Delta S° = \Sigma S°\text{ [products]} - \Sigma S°\text{ [reactants]}$

$= [(4 \times 211) + (6 \times 189)] - [(4 \times 193) + (5 \times 205)]$

$= 1978 - 1797$

$= +181\text{ JK}^{-1}$

$\Delta H° = \Sigma\Delta_f H°\text{[products]} - \Sigma\Delta_f H°\text{[reactants]}$

$= [(4 \times 90.4) + (6 \times -242)] - [(4 \times -46.2) + 0]$

$= -905.6\text{ kJ}$

$-905.6 - \left(298 \times \frac{181}{1000}\right)$

$\Delta G° = \Delta H° = T\Delta S° = -959.6\text{ kJ}$

(b) The reaction is spontaneous because ΔG is negative.

A closer look at $\Delta G = \Delta H - T\Delta S$

We can apply the equation to two general cases.

Reactions with a negative entropy change

For example, $N_2(g) + 3H_2(g) \rightarrow 2NH_3(g)$:

- ΔS is negative, and so $T\Delta S$ will also be negative.
- As temperature increases, $T\Delta S$ becomes more negative.
- $\Delta H - T\Delta S$ therefore becomes more positive as temperature increases.
- ΔG therefore becomes more positive.
- So a temperature must be reached at which the reaction is no longer feasible.

Reactions with a positive entropy change

For example, $H_2O(g) \rightarrow H_2(g) + \frac{1}{2}O_2(g)$:

- ΔS is positive, and so $T\Delta S$ will also be positive.
- As temperature increases, $T\Delta S$ becomes more positive.
- $\Delta H - T\Delta S$ therefore becomes more negative as temperature increases.
- ΔG therefore becomes more negative.
- So the reaction remains feasible as the temperature increases.

Exam practice

1 (a) Define the term 'lattice formation enthalpy'. [2]

(b) Write equations to show the lattice formation enthalpy of:

(i) lithium oxide [2]

(ii) calcium chloride [2]

(c) Write these ionic compounds in order of increasing lattice dissociation enthalpy — lowest first: NaCl, CsCl, KCl, LiCl
Explain the order that you choose. [3]

(d) Draw a Born–Haber cycle for the formation of sodium iodide. Label the species present at each stage. [3]

(e) Using your Born–Haber cycle from part (d) and the data given below, calculate a value for the enthalpy of formation of sodium iodide. [2]
Data/kJ mol^{-1}:

$\Delta_{latt}H$[NaI(s)] = −684 $\Delta_{at}H$[Na(s)] = +109 $\Delta_{i.e.}H$[Na(g)] = +494
$\Delta_{at}H$[iodine(g)] = +107 $\Delta_{e.a.}H$[I(g)] = −314

2 Carbon monoxide, CO, reacts with hydrogen gas, H_2, under certain conditions to form methanol, CH_3OH:

	$CO(g)$	+ $2H_2(g)$	→ $CH_3OH(l)$
S/J K^{-1} mol^{-1}	198	131	127
Δ_fH/kJ mol^{-1}	−111	0	−239

(a) Calculate the entropy change for the reaction. [2]

(b) Calculate the enthalpy change for the reaction. [2]

(c) Determine the Gibbs free energy change for the reaction at 298 K. [2]

(d) Given that the reaction is feasible at 298 K, show that the reaction ceases to be feasible at about 111°C. [2]

(e) Explain why a catalyst is added to the reaction mixture when it is used industrially. [1]

Answers and quick quiz 15 online

ONLINE

Summary

You should now have an understanding of:

- the important definitions used in thermodynamics
- the Born–Haber cycle and how to construct one
- the use of lattice enthalpy (formation or dissociation) as a measure of the attractive forces between ions
- how lattice enthalpy yields information about the nature of bonding types
- how lattice enthalpy, enthalpy of solution and hydration enthalpies are related in an energy cycle
- bond enthalpies and how they are used for calculating enthalpy changes
- how bond enthalpies can either be mean values or specific values
- the meaning of entropy and how to assess the entropy change in a physical or chemical process qualitatively
- how to calculate entropy changes using absolute entropy values
- how to calculate Gibbs free energy changes using $\Delta G = \Delta H - T\Delta S$
- the significance of Gibbs free energy in terms indicating whether reactions are feasible or not
- how to determine the temperature at which a reaction may, or may not, be feasible

16 Kinetics

Rate equations

What is meant by 'rate of a chemical reaction'?

REVISED

For the reaction:

A + B → X + Y

the **rate of reaction** is equal to the rate at which the concentration of reactant A or B is decreasing, or the rate at which the concentration of product X or Y is increasing.

Rate of reaction is defined as the rate of change of concentration of a reactant or a product with time. A rate of reaction has units mol dm^{-3} s^{-1}.

The rate of a reaction can be measured at any time during the reaction, but it is most conveniently done at the very start of the reaction, at time $t = 0$. This rate is called the **initial rate**. The initial rate of reaction is calculated by drawing the **tangent** to the concentration–time graph at $t = 0$ (Figure 16.1).

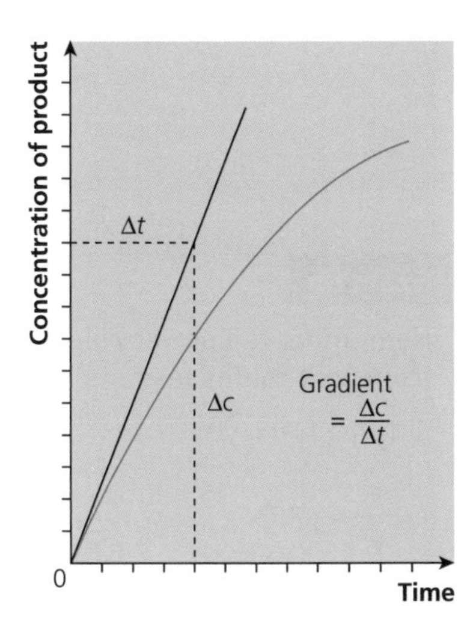

Figure 16.1 Calculating the initial rate of reaction

It is possible to measure the rate at any time during the reaction by measuring the gradient of the tangent at that particular time.

How does concentration affect the rate of a chemical reaction?

REVISED

For the general reaction between reactants A and B in solution:

A(aq) + B(aq) → X + Y

rate = k[A]^a[B]^b

where square brackets, [ ], represent the molar concentrations.

The variables a and b are called the **orders** of reaction, with respect to each reactant, and k is called the **rate constant**.

The rate constant is constant with time and varying concentration, but it increases with increasing temperature.

In the equation above, the order with respect to A is *a*, and the order with respect to B is *b*. Orders will only have the (integer) values 0, 1 and 2 in examples at this level.

Example

For the reaction in aqueous solution between propanone, $CH_3COCH_3(aq)$, and iodine, $I_2(aq)$, in the presence of an acid catalyst:

$$CH_3COCH_3 + I_2 \rightarrow CH_3COCH_2I + HI$$

the rate equation is found by experiment to be:

$$\text{rate} = k[CH_3COCH_3]^1[I_2]^0[H^+]^1$$

So the rate of this reaction is **first order** with respect to propanone, **first order** with respect to hydrogen ions and **zero order** with respect to iodine.

Typical mistakes

Many students think that a rate equation can be deduced from the balanced symbol equation. This is not true. It has to be determined experimentally.

The **overall order** of a reaction is simply the sum of all of the individual orders. If the rate equation is:

$$\text{rate} = k[CH_3COCH_3(aq)][H^+(aq)]$$

the overall order is 1 + 1 = 2, or **second order** overall.

The **units of the rate constant** can be found by rearranging the rate equation. For the propanone/iodine reaction above:

$$\text{rate} = k[CH_3COCH_3(aq)]^1[H^+(aq)]^1$$

Making *k* the subject of this equation gives:

$$k = \frac{\text{rate}}{[CH_3COCH_3(aq)]^1[H^+(aq)]^1}$$

So the units of *k* are:

$$\frac{\text{mol dm}^{-3}\,\text{s}^{-1}}{(\text{mol dm}^{-3})^2}$$

$$= \text{mol}^{-1}\,\text{dm}^3\,\text{s}^{-1}$$

Exam tip

Remember the power rules from your maths lessons:

$$(y^a) \times (y^b) = y^{(a+b)}$$

$$\frac{y^a}{y^b} = y^{(a-b)}$$

$$(y^a)^b = y^{ab}$$

Now test yourself

TESTED

1 The rate equation for the reaction:

$$(CH_3)_3CBr + OH^- \rightarrow (CH_3)_3COH + Br^-$$

is found to be rate = $k[(CH_3)_3CBr]$. The rate is measured in $\text{mol dm}^{-3}\,\text{s}^{-1}$.

(a) What are the orders with respect to $(CH_3)_3CBr$ and OH^-?

(b) What is the overall order of the reaction?

(c) Give the units of the rate constant.

Answers on p. 220

How do concentration and rate change with time in a reaction?

REVISED

When the concentration of a reactant in a chemical reaction is monitored with time, it can vary in one of three main ways, as indicated by the graphs in Figure 16.2.

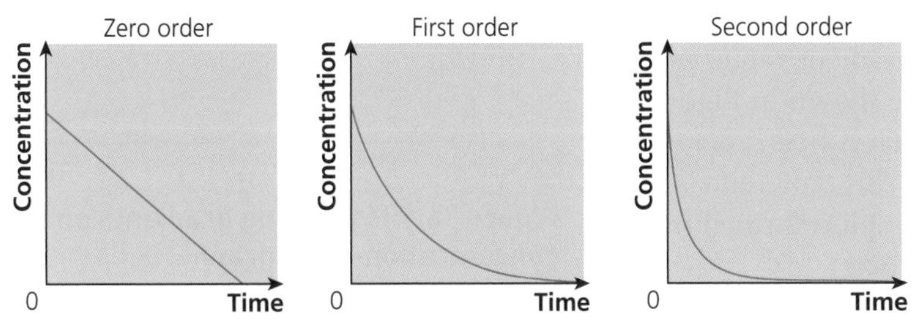

Figure 16.2 Concentration–time graphs

When the rates of reaction are monitored against concentration, the graphs shown in Figure 16.3 are obtained.

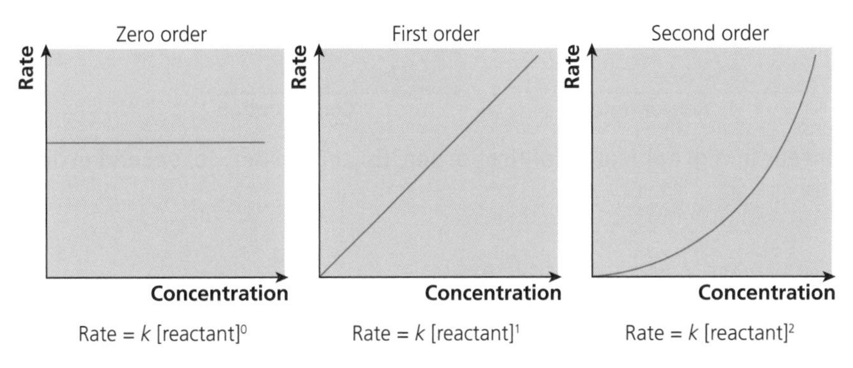

Figure 16.3 Rate–concentration graphs

- Zero order — as the concentration is doubled, the rate is unaffected.
- First order — as the concentration is doubled, the rate increases by a factor of 2.
- Second order — as the concentration is doubled, the rate increases by a factor of $2^2 = 4$.

> **Exam tip**
>
> Remember — 'first order' means that when the concentration of the reactant is multiplied by a factor x, the rate is also multiplied by the factor x.

Determining orders of reactions using experimental data

REVISED

Using graphical methods

If a graph is plotted of the concentration of a reactant against time, the order of the reaction can be deduced using that graph.

The easy way of determining whether a reaction is first order or not is to measure the half-life time —that is, the time that elapses when half of the concentration of a reactant remains, and then if this is halved again etc. (Figure 16.4).

If the half lives are the same, then the reaction is first order.

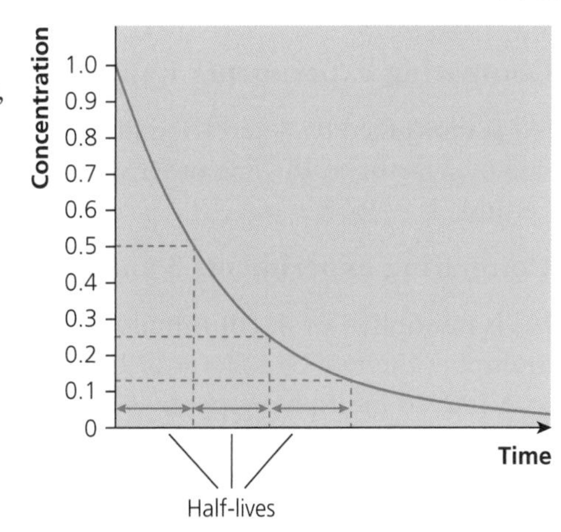

Figure 16.4 Measuring half-life

Another method to determine the order is to draw tangents to the curve at various points, and then measure their gradients, as shown in Figure 16.5. The gradients (as positive values) are equal to the rate of reaction at that time.

So a list is then drawn up of different rates (using the tangent values) and the corresponding concentration values from the same graph.

If another graph is now plotted of rate (the gradient value) against the concentration, a straight line like the one shown in Figure 16.6a will be produced if the order is one, that is first order. If the graph produced is a flat, horizontal line and this represents zero order (Figure 16.6b). If a parabola ($y = x^2$) graph is formed then this means that it is second order (Figure 16.6c).

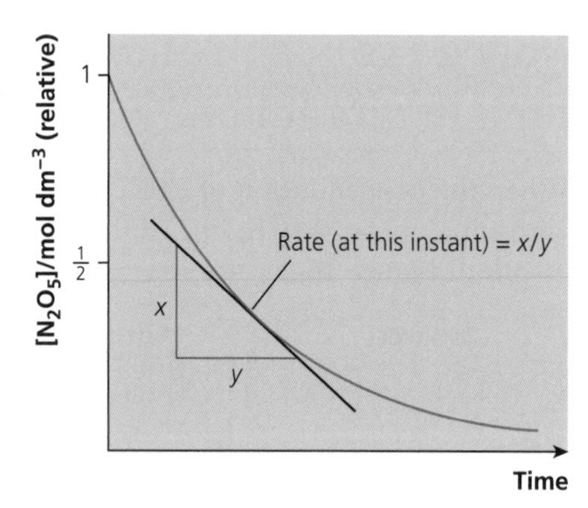

Figure 16.5 Measuring gradients on a concentration–time graph

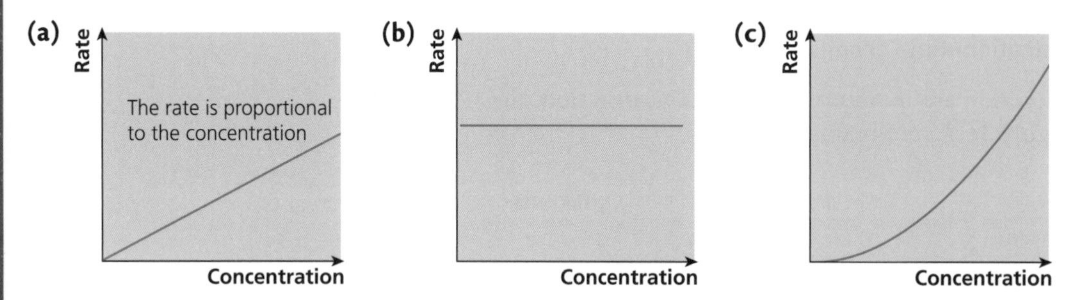

Figure 16.6 Determining the order of reaction graphically: (a) first order; (b) zero order; (c) second order

Using initial rates

In a reaction between reactants A, B and C, the following initial rates were obtained using the initial concentrations shown in four different experiments.

	Initial concentration/mol dm^{-3}			
Experiment	**[A]**	**[B]**	**[C]**	**Initial rate/mol dm^{-3} s^{-1}**
1	0.05	0.03	0.12	1.20×10^{-4}
2	0.20	0.06	0.24	3.84×10^{-3}
3	0.05	0.03	0.24	2.40×10^{-4}
4	0.20	0.03	0.12	1.92×10^{-3}

Comparing experiments 1 and 3:

[A] and [B] remain the same but [C] doubles. This doubles the rate, so the order with respect to C must be 1.

Comparing experiments 1 and 4:

[A] is multiplied by 4 and [B] and [C] stay constant. This multiplies the rate by a factor of 16. The order with respect to A must therefore be 2, because $4^2 = 16$.

Comparing experiments 3 and 2:

[A] is multiplied by 4, [B] is multiplied by 2 and [C] stays constant. This multiplies the rate by a factor of 16. We know that the order with respect to A is 2, so multiplying [A] by 4 should multiply the rate by 16 — and it does. Therefore, the change in [B] has no effect on the rate. So the order with respect to B must be 0.

The overall rate equation is therefore:

rate = $k[A]^2[B]^0[C]^1$ or = $k[A]^2[C]^1$

and the reaction is third order overall.

The value for the rate constant can be calculated using any of experiments 1, 2, 3 or 4. Using the data from experiment 1 and that rate = $k[A]^2[C]$ gives:

$$1.2 \times 10^{-4} = k \times 0.05^2 \times 0.03^0 \times 0.12^1$$

This gives $k = 0.4$, with units:

$$\frac{\text{mol dm}^{-3}\,\text{s}^{-1}}{(\text{mol dm}^{-3})^3}$$

$= \text{mol}^{-2}\,\text{dm}^6\,\text{s}^{-1}$

Now test yourself

TESTED

2 The following data were measured for the reaction between nitrogen(II) oxide and hydrogen:

$2NO(g) + 2H_2(g) \rightarrow N_2(g) + 2H_2O(g)$

Experiment number	Initial concentrations/mol dm^{-3}		Initial rate/mol dm^{-3} s^{-1}
	[NO(g)]	[H_2(g)]	
1	0.100	0.100	1.11×10^{-3}
2	0.100	0.200	2.24×10^{-3}
3	0.200	0.100	4.45×10^{-3}

(a) Determine the order with respect to NO and the order with respect to H_2.
(b) What is the overall order of reaction?
(c) Write the rate equation for the reaction.
(d) Determine a value for the rate constant and deduce its units.

Answers on p. 220

The effect of temperature on rate of reaction

In all chemical reactions, as the temperature is increased the rate will increase.

A typical Maxwell–Boltzmann distribution is shown in Figure 16.7.

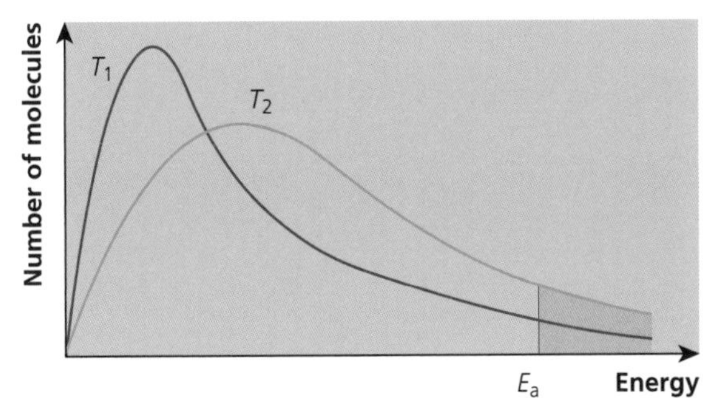

Figure 16.7 Graph of a Maxwell–Boltzmann distribution at two different temperatures, where $T_2 > T_1$

The rate increases with increasing temperature because the proportion of particles having energy greater than the activation energy increases. This increases the number of successful collisions taking place per unit time.

Given a typical rate equation:

$$\text{rate} = k[A]^2[C]$$

as the temperature increases, the value of k will also increase.

- The rate constant is related to the shaded area of the Maxwell–Boltzmann distribution in Figure 16.7 to the right of the activation energy value. The larger this area, the more particles have energy $E > E_a$ so both the rate and the rate constant will be larger.
- The lower the temperature, the smaller will be the proportion of molecules having an energy greater than the activation energy. Therefore, the rate constant is smaller.

The Arrhenius equation

REVISED

The Arrhenius equation relates the rate constant to the temperature for a reaction, and is shown in Figure 16.8.

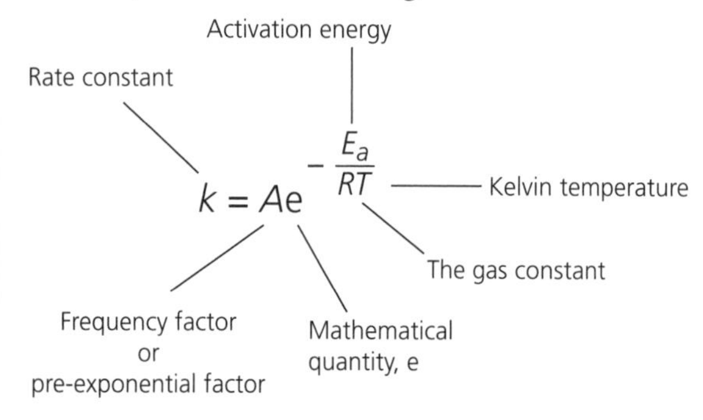

Figure 16.8 The Arrhenius equation

Using this equation, it is possible to determine the activation energy, E_a, for a process, and also determine the value of the pre-exponential factor, A.

If natural logs are taken of both sides of the Arrhenius equation, the following relationship results:

$$\ln k = \frac{-E_a}{RT} + \ln A$$

This is in the form $y = mx + c$, where m is the gradient and c is the y-axis intercept. But in this case, $\left(\frac{1}{T}\right)$ is the x term, and $\ln A$ is the c term. So if a graph of $\ln k$ (the natural log of the rate constant) is plotted against $\left(\frac{1}{T}\right)$, a straight line should be formed, of gradient $\left(\frac{-E_a}{R}\right)$ and the intercept on the y-axis will be $\ln A$ (Figure 16.9).

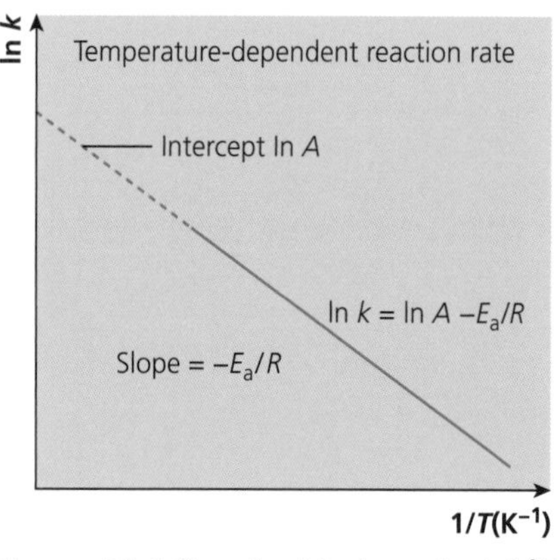

Figure 16.9 Graph of $\ln k$ against $1/T$

Once the gradient has been determined using the graph, this gradient is then equal to $\left(\frac{-E_a}{R}\right)$.

This gives:

$$R = \frac{-E_a}{\text{gradient of line}}$$

So:

$$E_a = -\text{gradient of line} \times R$$

The activation energy is then determined with units of $kJ\,mol^{-1}$.

Rate equations and reaction mechanism

A rate equation tells us how the concentrations of the reactants affect the rate of the reaction. However, it can also indicate something about the **mechanism** of the reaction taking place. For example, nitrogen(IV) oxide reacts with fluorine according to the equation:

$$2NO_2(g) + F_2(g) \rightarrow 2NO_2F(g)$$

The rate equation, found by experiment, is first order with respect to both reactants:

$$\text{rate} = k[NO_2(g)][F_2(g)]$$

We can deduce the following about the mechanism of the reaction:

- Since both NO_2 and F_2 are in the rate equation as non-zero orders, this means that both concentrations affect the rate.
- Any species that is zero order in the mechanism must be involved in a subsequent fast step.

Therefore, it can be deduced that:

- NO_2 and F_2 are involved in the **rate-determining step** of the mechanism — so the first step could be:

$$NO_2(g) + F_2(g) \rightarrow NO_2F(g) + F(g) \qquad \text{slow}$$

- The next step — a fast one — may involve F(g) from the slow step reacting with another NO_2 molecule:

$$NO_2 + F \rightarrow NO_2F \qquad \text{fast}$$

Notice that the two individual steps in this mechanism add up to give the overall chemical equation:

$$NO_2(g) + F_2(g) \rightarrow NO_2F(g) + F(g) \quad (1)$$

$$NO_2(g) + F(g) \rightarrow NO_2F(g) \quad (2)$$

$$2NO_2(g) + F_2(g) \rightarrow 2NO_2F(g)$$

The species appearing in the rate equation also occur in the **rate-determining step** of the mechanism.

Exam practice

1 In a reaction taking place between substances A and B, the following results were obtained in three different experiments.
Assume that the concentrations given in the table are all initial concentrations.

	[A] $mol\,dm^{-3}$	[B] $mol\,dm^{-3}$	Initial rate $mol\,dm^{-3}\,s^{-1}$
Experiment 1	1.0×10^{-2}	4.0×10^{-3}	3.20×10^{-3}
Experiment 2	1.0×10^{-2}	8.0×10^{-3}	1.28×10^{-2}
Experiment 3	2.0×10^{-2}	8.0×10^{-3}	2.56×10^{-2}

(a) Deduce the order of the reaction with respect to A, and the order of the reaction with respect to B. [2]
(b) Hence, write the rate equation for the reaction and calculate the rate constant, stating its units. [3]
(c) Calculate the rate of reaction when the initial concentrations of A and B are both $6.0 \times 10^{-3}\,mol\,dm^{-3}$. [1]
(d) In another experiment in which the initial concentrations of A and B are both $x\,mol\,dm^{-3}$, the initial rate of reaction is found to be $9.22 \times 10^{-2}\,mol\,dm^{-3}\,s^{-1}$. Calculate the value of x. [2]

2 Propanone, in acidic solution, reacts with iodine according to:

$CH_3COCH_3 + I_2 \rightarrow CH_3COCH_2I + HI$

In an experiment, the time taken for the iodine to reach a certain concentration was measured. The concentration of hydrogen ions was kept constant in all four experiments described in the table below. The order with respect to the acid is known to be 1.

$[CH_3COCH_3]$ $mol\,dm^{-3}$	$[I_2]$ $mol\,dm^{-3}$	Time s
0.25	0.05	68
0.50	0.05	34
1.00	0.05	17
0.50	0.10	34

(a) Deduce the order of reaction with respect to propanone. [1]
(b) Deduce the order with respect to iodine. [1]
(c) Hence, write the rate equation for the reaction. [1]
(d) Comment, with a reason, on whether or not the following mechanism is consistent with the rate equation you have suggested in part (c). [2]

Step 1: $CH_3COCH_3 + H^+ \rightarrow CH_3COH^+CH_3$ slow

Step 2: $CH_3COH^+CH_3 \rightarrow CH_3C(OH)CH_2 + H^+$ fast

Step 3: $CH_3C(OH)CH_2 + I_2 \rightarrow CH_3COCH_2I + HI$ fast

Answers and quick quiz 16 online

ONLINE

Summary

You should now have an understanding of:

- what is meant by rate of reaction and initial rate
- the effect of changes in concentration on rate
- a rate constant and how to determine its units
- rate equations and how to determine them given initial rate information
- concentration–time and rate–concentration graphs
- the effect of temperature on reaction rate
- the link between a rate equation and a reaction mechanism

17 Equilibria

The equilibrium constant, K_p

It is more convenient with gases to measure their partial pressures rather than their concentration:

$$\text{partial pressure, } p\text{, of a gas} = \frac{\text{number of moles of gas}}{\text{total number of moles}} \times \text{total pressure}$$

Or:

$$p_a = x_a P_{total}$$

where p_a is the partial pressure of the gas, x_a is the mole fraction of the gas and P_{total} is the total pressure.

Writing expressions for K_p

REVISED

These are similar to writing an expression for K_c, but square brackets denoting concentration are replaced with curved brackets instead.

For the reaction $N_2(g) + 3H_2(g) \rightleftharpoons 2NH_3(g)$:

$$K_p = \frac{p^2(NH_3(g))}{p(N_2(g)) \times p^3(H_2(g))}$$

Units will be $(atm)^2/(atm \times atm^3) = atm^{-2}$.

All partial pressures are measured at equilibrium.

Calculations involving K_p

REVISED

Example 1

Calculate the value of K_p for the reaction $N_2(g) + 3H_2(g) \rightleftharpoons 2NH_3(g)$, given that the number of moles at equilibrium of $N_2(g)$ = 1.2, $H_2(g)$ = 0.4 and $NH_3(g)$ = 2.3, and the total pressure is 6 atmospheres.

Answer

Start by calculating the partial pressures for each component:

total number of moles = 1.2 + 0.4 + 2.3 = 3.9 moles

	$pH_2(g)$	$pN_2(g)$	$pNH_3(g)$
Mole fractions:	$\frac{0.4}{3.9}$	$\frac{1.2}{3.9}$	$\frac{2.3}{3.9}$
Partial pressures:	$\frac{0.4}{3.9} \times 6$	$\frac{1.2}{3.9} \times 6$	$\frac{2.3}{3.9} \times 6$
	0.615 atm	1.846 atm	3.538 atm

Substituting into the expression for K_p:

$$K_p = \frac{p^2(NH_3(g))}{p(N_2(g))} \times p^3(H_2(g)) = \frac{(3.538)^2}{1.846 \times (0.615)^3} = 29.151\,atm^{-2}$$

Example 2

Calculate the value for K_p from the following information: 1 mole of hydrogen and 3 moles of nitrogen are mixed and allowed to come to equilibrium at 250 atmospheres. It is found, on analysing the reaction mixture, that 0.33 moles of hydrogen remain.

Answer

Write the balanced equation:	$N_2(g)$	+	$3H_2(g)$	$\rightleftharpoons$	$2NH_3(g)$
Starting moles:	1		3		0
Moles left at equilibrium:	$1 - x$		$3 - 3x$		$2x$

Where x is the number of moles of nitrogen that react. But the number of moles of hydrogen remaining is 0.33 and we know that the number of moles of nitrogen must be a third of this figure (they are mixed originally in the ratio 3:1 respectively). Therefore, there must be 0.11 mol of nitrogen remaining. Using the expression above, it can be seen that for nitrogen, the number of moles at equilibrium is 1 – x but 1 – x is equal to 0.11. Therefore, if $1 - x = 0.11$:

$x = 1 - 0.11 = 0.89$ mol

number of moles of ammonia, $2x = 2 \times 0.89 = 1.78$ mol

total number of moles = 0.11 + 0.33 + 1.78 = 2.22 mol

	N_2	H_2	NH_3
Calculate the mole fractions:	$\frac{0.11}{2.22}$	$\frac{0.33}{2.22}$	$\frac{1.78}{2.22}$
Convert to partial pressures:	$\frac{0.11}{2.22} \times 250$	$\frac{0.33}{2.22} \times 250$	$\frac{1.78}{2.22} \times 250$
	12.387 atm	37.162 atm	200.450 atm

Substituting into the expression for K_p:

$$K_p = \frac{p^2\,(NH_3(g))}{p\,(N_2(g))} \times p^3(H_2(g)) = \frac{(200.450)^2}{12.387 \times (37.162)^3} = 0.0632\ \text{atm}^{-2}$$

The effect of temperature on the equilibrium constant, K_p

REVISED

As with K_c, the only factor to affect K_p is temperature. All other variables, such as a catalyst, pressure or changing the concentrations or pressures of reactants or products, have no effect on the value of the equilibrium constant.

On relating the two equations for $\Delta G^\ominus$, that is:

$\Delta G^\ominus = \Delta H^\ominus - T\Delta S^\ominus$ and $\Delta G^\ominus = -RT\ln K$

we produce:

$\Delta H^\ominus - T\Delta S^\ominus = -RT\ln K$

where K could be either K_c or K_p.

Therefore, making $\ln K$ the subject gives:

$$\ln K = \frac{\Delta S^\ominus}{R} - \frac{\Delta H^\ominus}{RT}$$

Typical mistake

A catalyst has no effect on the composition of the resulting equilibrium mixture, so the value of K_p (or K_c) remains unchanged. A catalyst only affects the *rate* at which the eventual equilibrium mixture is formed — the composition is not changed.

So, if a graph is plotted of $\ln(K_c$ or $K_p)$ on the y-axis versus $1/T$ (where the temperature is measured in Kelvin), a straight line is obtained of gradient $-\Delta H^\ominus/R$ and the intercept on the y-axis is $\Delta S^\ominus/R$.

Figure 17.1 shows the type of straight-line relationship expected for (a) an exothermic reaction and (b) an endothermic reaction.

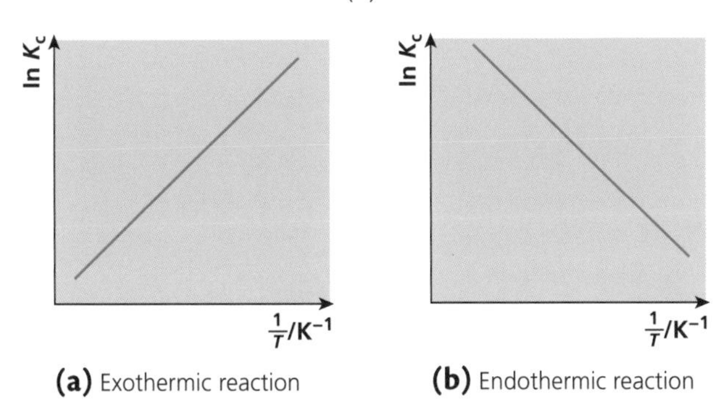

(a) Exothermic reaction **(b)** Endothermic reaction

Figure 17.1 Graphs for exothermic and endothermic reactions

The sign of the gradient in each case is determined by the sign of the enthalpy change (it will be positive for exothermic reactions and negative for endothermic reactions) because the gradient of the line is equal to $-\Delta H^\ominus/R$.

For exothermic reactions it can be seen that, on increasing the temperature, $1/T$ decreases and $\ln K$ becomes smaller. In other words, an increase in temperature reduces the equilibrium constant. This means that there is a higher proportion of reactants and less product in the equilibrium mixture as the temperature increases (Le Chatelier's principle also predicts this — see p. 53). The opposite is true for the effect of temperature on endothermic reactions. The influence of temperature on the value of K_p for two reactions is shown in Tables 17.1 and 17.2.

Table 17.1 An exothermic reaction: $N_2(g) + 3H_2(g) \rightleftharpoons 2NH_3(g)$

T/K	K_p/atm^{-2}
400	1.0×10^{2}
500	1.6×10^{-1}
600	3.1×10^{-3}
700	6.3×10^{-5}
K_p gets smaller with increasing temperature	

Table 17.2 An endothermic reaction: $N_2O_4(g) \rightleftharpoons 2NO_2(g)$

T/K	K_p/atm^{-2}
275	2.2×10^{-2}
350	4.5
500	1.5×10^{3}
K_p gets larger with increasing temperature	

You should be able to plot a graph of the natural log (ln) of the equilibrium constant ($\ln K$) versus the reciprocal of temperature ($1/T$) and use it to determine the value for $\Delta H^\ominus$ or $\Delta S^\ominus$.

Now test yourself

TESTED

1 The equilibrium constant for the equilibrium reaction:

$N_2H_4(g) \rightleftharpoons N_2(g) + 2H_2(g)$

varies with temperature as shown in the graph.

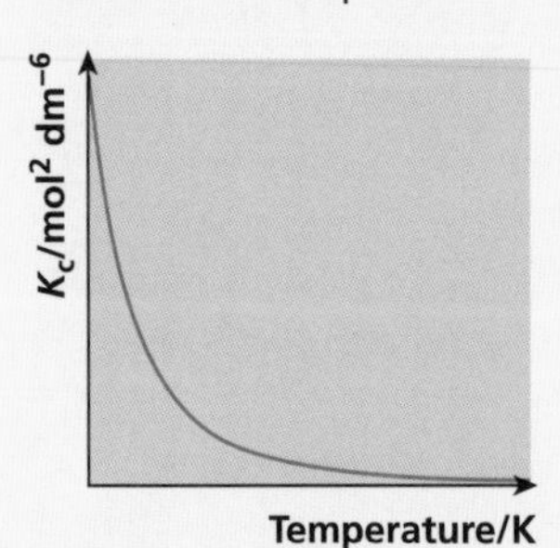

(a) What happens to the value of the equilibrium constant as the temperature increases?
(b) Is the forward reaction in the equilibrium exothermic or endothermic? Explain your answer.
(c) Given the data in the graph, explain how another graph could be plotted that is a straight line.
(d) Using your graph from part (c), explain how the following can be determined:
  (i) the enthalpy change for the reaction
  (ii) its entropy change.

Answers on p. 220

Exam practice

1 A gaseous equilibrium is established by adding 2.00 mol of sulfur(IV) oxide to 2.00 mol of oxygen at 200 atmospheres and waiting for equilibrium to be attained:

$2SO_2(g) + O_2(g) \rightleftharpoons 2SO_3(g)$

It is found that 0.550 mol of sulfur(VI) oxide is present in the equilibrium mixture. The total volume of the reaction mixture is 10.0 dm^3.

(a) Calculate the amounts, in moles, of SO_2 and O_2 at equilibrium. [2]
(b) Calculate the concentrations of all species taking part in the reaction. [3]
(c) Write an expression for K_c for the reaction. [1]
(d) Hence determine the value of K_c for the equilibrium, stating the units. [3]

Sulfur dioxide, SO_2, is added to the equilibrium, and the pressure is increased at constant volume (maintaining the same temperature).

(e) State and explain the effect on:
(i) the position of equilibrium [2]
(ii) the value of K_c [2]

Answers and quick quiz 17 online

ONLINE

Summary

You should now have an understanding of:

- the equilibrium constant, K_p
- the quantitative effects of changes of temperature in terms of the position of equilibrium and on K_p
- how to perform various calculations using K_p

18 Redox equilibria

Redox equations

Many reactions are classed as **redox reactions** because they involve both **reduction** and **oxidation** of different species.

Reduction is the gain of electrons.

Oxidation is the loss of electrons.

Exam tip

Oxidation can also be defined as the gain of oxygen and loss of hydrogen. Reduction is the loss of oxygen and gain of hydrogen.

Half-equations

REVISED

In any redox reaction, one species is oxidised and another is reduced. We can write two separate equations to show what is going on in each 'half' of the reaction in terms of electrons added or removed.

Example

When an acidified solution containing manganate(VII) ions, $MnO_4^-(aq)$, is added to aqueous iron(II) ions, a reaction occurs in which manganese(II) ions and iron(III) ions are formed. Write the two half-equations for the redox reactions taking place.

Answer

Oxidation: $Fe^{2+}(aq) \rightarrow Fe^{3+}(aq) + e^-$

For the reduction, put the equation together according to these steps:

- write down what is known: $MnO_4^- \rightarrow Mn^{2+}$
- add water, H_2O, to balance the oxygens: $MnO_4^- \rightarrow Mn^{2+} + 4H_2O$
- add hydrogen ions, H^+, to balance the hydrogens:
  $MnO_4^- + 8H^+ \rightarrow Mn^{2+} + 4H_2O$
- add electrons, e^-, to balance the charge:
  $MnO_4^- + 8H^+ + 5e^- \rightarrow Mn^{2+} + 4H_2O$

The two half-equations can then be added together to form an overall equation, after making sure that the electrons cancel out:

Oxidation: $Fe^{2+}(aq) \rightarrow Fe^{3+}(aq) + e^-$

Reduction: $MnO_4^- + 8H^+ + 5e^- \rightarrow Mn^{2+} + 4H_2O$

For the overall balanced equation, multiply the top half-equation by 5 and then add it to the bottom half-equation:

$5Fe^{2+}(aq) \rightarrow 5Fe^{3+}(aq) + 5e^-$

$MnO_4^- + 8H^+ + 5e^- \rightarrow Mn^{2+} + 4H_2O$

$MnO_4^-(aq) + 8H^+(aq) + 5Fe^{2+}(aq) \rightarrow Mn^{2+}(aq) + 5Fe^{3+}(aq) + 4H_2O(l)$

Exam tip

The numbers of electrons in each half-equation must be the same before they are added together.

Oxidation states

REVISED

The **oxidation state** of an element can be thought of as the 'formal' charge that the element has when combined in a compound. For example, in $CrCl_3$ the oxidation state of the chlorine is −1 and that of the chromium is +3 (the sum of the individual oxidation states is zero for a compound).

Transition metal ions can often exist in more than one oxidation state. We must indicate in the name of a compound just what the oxidation state of the metal is. For example, we use names such as copper(II) sulfate (with Cu^{2+} ions), copper(I) oxide (with Cu^{+} ions) and chromium(III) fluoride (with Cr^{3+} ions).

If an **oxidation state** increases in a reaction, then oxidation has occurred (loss of electrons) and reduction has occurred if an oxidation state has decreased.

Now test yourself

TESTED

1 When dichromate(VI) ions, $Cr_2O_7^{2-}$(aq), are added to an acidified solution containing iodide ions, I^-(aq), a redox reaction occurs in which an aqueous solution containing chromium(III) ions and iodine is formed.
   (a) Write the two half-equations for the reaction.
   (b) Which substances have been (i) reduced and (ii) oxidised in the reaction?
   (c) Write the overall equation for the reaction by combining the two half-equations from part (a).
   (d) What are the oxidation state changes of (i) chromium and (ii) iodine in the reaction?

2 A solution containing the ferrate(VI) ion, FeO_4^{2-}(aq) is added to an acidified solution containing manganese(IV) oxide powder. An aqueous solution containing iron(III) ions and manganate(VII) ions is formed.
   (a) Write the two half-equations for the reaction.
   (b) Which substance has been (i) reduced and (ii) oxidised in the reaction?
   (c) Write the overall equation by combining the two half-equations from part (a).
   (d) What are the oxidation state changes of (i) manganese and (ii) iron in the reaction?

Answers on p. 220

Electrode potentials

Measuring a standard electrode potential

REVISED

A **standard reference electrode** — a hydrogen electrode — is used to measure a **standard electrode potential**. The potential difference of this electrode under standard conditions is defined as 0.00 volts.

Figure 18.1 shows how the standard electrode potential is measured for the $Cu^{2+}(aq) + 2e^- \rightleftharpoons Cu(s)$ system. Note that half-equations for electrode systems are always written as reductions.

The **standard electrode potential** of a substance is the potential difference, in volts, of a substance in an aqueous solution of its ions relative to the standard hydrogen electrode, under standard conditions: 298 K, 100 kPa and 1 mol dm^{-3} solutions.

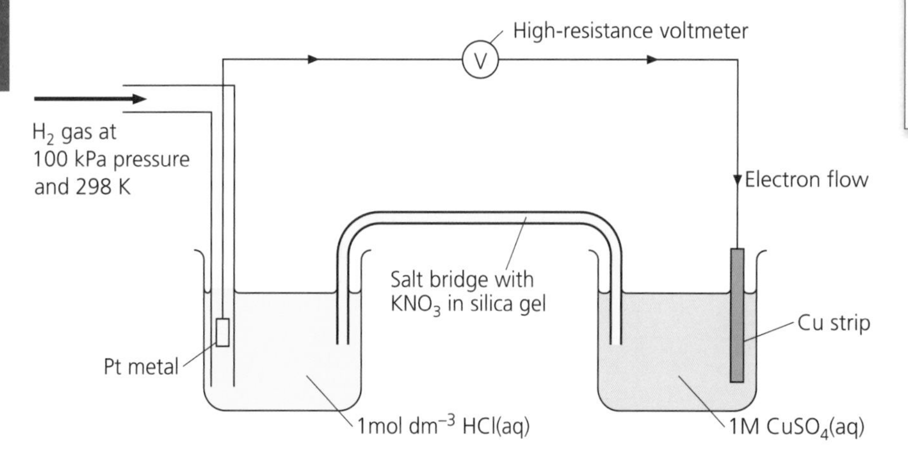

Figure 18.1 Measuring an electrode potential

Two half-cells are shown — the one on the left is a standard hydrogen electrode and the one on the right shows a piece of copper in contact with Cu^{2+} ions. A high-resistance voltmeter is used so that it reads the potential difference with, effectively, no current passing through it.

A reading of +0.34 volts is recorded from the cell terminals, as measured in Figure 18.1. So we say that the electrode potential for the $Cu^{2+}(aq) + 2e^- \rightleftharpoons Cu(s)$ system is +0.34 V under standard conditions.

The two half-equations (written as reduction processes) are:

Hydrogen electrode: $2H^+(aq) + 2e^- \rightleftharpoons H_2(g)$ $E^\ominus = 0.00\,V$

Copper electrode: $Cu^{2+}(aq) + 2e^- \rightleftharpoons Cu(s)$ $E^\ominus = +0.34\,V$

Some points to note:

- $E^\ominus$ is used to indicate an electrode potential measured under standard conditions.
- The more positive electrode potential shifts to the right-hand side:

  $Cu^{2+}(aq) + 2e^- \rightarrow Cu(s)$

- The more negative electrode potential shifts to the left-hand side:

  $H_2(g) \rightarrow 2H^+(aq) + 2e^-$

- Electrons are gained by the copper(II) ions at the copper electrode, which will therefore have a positive charge. Reduction occurs at this electrode.
- Electrons are released at the hydrogen electrode, which will therefore have a negative charge. Oxidation takes place at this electrode.
- Electrons move from the hydrogen half-cell to the copper half-cell in the external circuit — from the negative terminal to the positive terminal.
- Adding the two half-equations together gives the overall equation:

  $Cu^{2+}(aq) + H_2(g) \rightarrow Cu(s) + 2H^+(aq)$

- The overall electrode potential is equal to the two individual processes added together, but the reverse process has its sign changed:

  $E^\ominus = +0.34\,V + 0.00\,V = +0.34\,V$

- Because the overall electrode potential is positive, this means that the reaction as written is spontaneous, that is:

  $Cu^{2+}(aq) + H_2(g) \rightarrow Cu(s) + 2H^+(aq)$

Example

Some reduction processes, together with their electrode potentials, are written below:

(1) $Zn^{2+}(aq) + 2e^- \rightleftharpoons Zn(s)$ $E^\ominus = -0.76\,V$

(2) $I_2(aq) + 2e^- \rightleftharpoons 2I^-(aq)$ $E^\ominus = +0.54\,V$

(3) $Fe^{3+}(aq) + e^- \rightleftharpoons Fe^{2+}(aq)$ $E^\ominus = 0.77\,V$

Combine the following pairs of equations, giving the overall electrode potentials for the spontaneous processes:

(a) (1) and (2)

(b) (2) and (3)

Typical mistake

If asked to draw a diagram showing how the standard electrode potential for $Fe^{3+}(aq) + e^- \rightleftharpoons Fe^{2+}(aq)$ is measured, it is important to include 1.00 mol dm^{-3} solutions of *both* Fe^{2+} ions *and* Fe^{3+} ions in contact with a platinum (inert) electrode.

Answer

(a) Equilibrium (2) shifts to the right-hand side because it is more positive; equilibrium (1) shifts to the left-hand side:

(1) $Zn^{2+}(aq) + 2e^- \rightleftharpoons Zn(s)$; $E^\ominus = -0.76$ V (shifts left, ←)

(2) $I_2(aq) + 2e^- \rightleftharpoons 2I^-(aq)$; $E^\ominus = +0.54$ V (shifts right, →)

Reversing equilibrium 2 and adding gives:

$$Zn(s) + I_2(aq) \rightarrow Zn^{2+}(aq) + 2I^-(aq)$$

The overall electrode potential for this spontaneous reaction will be:

$E^\ominus = +0.54 + (+0.76)$

$= +1.30$ V

The sign of the electrode potential is positive, so the reaction, in which zinc reduces iodine to iodide ions and zinc atoms are oxidised to zinc ions, is spontaneous.

(b) Equilibrium 3 shifts to the right-hand side — it is more positive; equilibrium 2 shifts to the left-hand side:

(2) $I_2(aq) + 2e^- \rightleftharpoons 2I^-(aq)$; $E^\ominus = +0.54$ V (shifts left, ←)

(3) $Fe^{3+}(aq) + e^- \rightleftharpoons Fe^{2+}(aq)$; $E^\ominus = +0.77$ V (shifts right, →)

Reversing equilibrium 2 and multiplying equilibrium 3 by 2 to balance the electrons and then adding gives:

$$2Fe^{3+}(aq) + 2I^-(aq) \rightarrow 2Fe^{2+}(aq) + I_2(aq)$$

The overall electrode potential for this reaction, the spontaneous one, will be:

$E^\ominus = +0.77 + (-0.54)$

$= +0.23$ V

So, iron(III) ions will oxidise iodide ions to form iodine and iron(II) ions.

Now test yourself

TESTED

3 Predict the products of the reactions, if any, that occur when aqueous iron(III) ions are added to solutions containing chloride ions, bromide ions and iodide ions in three separate experiments. Use the following electrode potentials:

$Cl_2(aq) + 2e^- \rightleftharpoons 2Cl^-(aq)$ $E^\ominus = +1.36$ V

$Br_2(aq) + 2e^- \rightleftharpoons 2Br^-(aq)$ $E^\ominus = +1.07$ V

$I_2(aq) + 2e^- \rightleftharpoons 2I^-(aq)$ $E^\ominus = +0.54$ V

$Fe^{3+}(aq) + e^- \rightleftharpoons Fe^{2+}(aq)$ $E^\ominus = +0.77$ V

Answers on p. 220

Electrochemical series

Standard electrode potentials can be written as a list called the **electrochemical series**, in which each electrochemical process is written as a reduction process. Table 18.1 gives an example of an electrochemical series.

Table 18.1 An electrochemical series

Oxidising power of left-hand species increases ↑		$E^\ominus$/V	Reducing power of right-hand species increases ↓
	$F_2(g) + 2e^- \rightarrow 2F^-(aq)$	+2.87	
	$H_2O_2(aq) + 2H^+(aq) + 2e^- \rightarrow 2H_2O(1)$	+1.77	
	$Au^+(aq) + e^- \rightarrow Au(s)$	+1.68	
	$Cl_2(g) + 2e^- \rightarrow 2Cl^-(aq)$	+1.36	
	$O_2(g) + 4H^+(aq) + 4e^- \rightarrow 2H_2O(1)$	+1.23	
	$Br_2(l) + 2e^- \rightarrow 2Br^-(aq)$	+1.09	
	$Ag^+(aq) + e^- \rightarrow Ag(s)$	+0.80	
	$Fe^{3+}(aq) + e^- \rightarrow Fe^{2+}(aq)$	+0.77	
	$I_2(s) + 2e^- \rightarrow 2I^-(aq)$	+0.54	
	$O_2(g) + 2H_2O(1) + 4e^- \rightarrow 4OH^-(aq)$	+0.40	
	$Cu^{2+}(aq) + 2e^- \rightarrow Cu(s)$	+0.34	
	$S(s) + 2H^+(aq) + 2e^- \rightarrow H_2S(g)$	+0.14	
	$2H^+(aq) + 2e^- \rightarrow H_2(g)$	0.00	
	$Pb^{2+}(aq) + 2e^- \rightarrow Pb(s)$	−0.13	
	$Sn^{2+}(aq) + 2e^- \rightarrow Sn(s)$	−0.14	
	$Ni^{2+}(aq) + 2e^- \rightarrow Ni(s)$	−0.25	
	$Co^{2+}(aq) + 2e^- \rightarrow Co(s)$	−0.28	

Notice the following about this series:

- Species on the left are oxidising agents.
- Species on the right are reducing agents.
- The electrode potentials increase in positive value going up the list.
- This means that the oxidising power of the species on the left of the equilibrium at the top of the list (F_2) is greater than all those below it.
- An oxidising agent at the top of the list (on the left) will oxidise any reducing agent below it (on the right). For example, fluorine gas, $F_2(g)$, will oxidise anything below it on the right-hand side, such as $Br^-(aq)$, $Ag(s)$, $Fe^{2+}(aq)$, $I^-(aq)$, $OH^-(aq)$, $Cu(s)$ and $H_2S(s)$.
- The reducing power of the species on the right increases going down the list.
- Species like Co(s), Ni(s), Sn(s), Pb(s) will all reduce any species on the left-hand side above them in the list.

Representation of electrochemical cells

REVISED

For all electrochemical cells (combinations of two half-cells), a **conventional representation** can be written that includes the redox processes taking place in the cell. This cell 'diagram' is a symbolic form, representing the electrochemical features and processes taking place. Figure 18.2 shows this conventional representation.

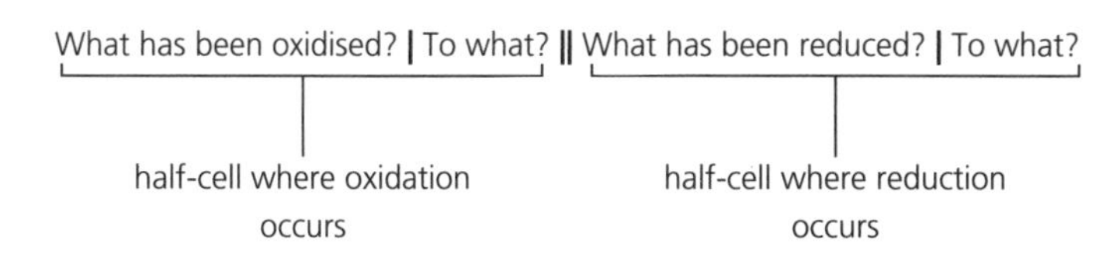

Figure 18.2 Representing an electrochemical cell

The vertical straight lines represent phase-change boundaries — for example, a solid in contact with a solution. The double vertical line represents the salt bridge through which ions flow.

Calculating overall cell potentials

A cell potential is also called an electromotive force (e.m.f.). An overall cell potential, $E^\ominus_{cell}$, is calculated using:

$$E^\ominus_{cell} = E^\ominus_{rhs} - E^\ominus_{lhs}$$

where $E^\ominus_{rhs}$ and $E^\ominus_{lhs}$ are the two electrode potentials for the half-equations written as reductions.

Exam tip

If an overall electrode potential is positive then the redox process being considered is feasible but the rate is not predictable.

Example

Consider the two half-equations written as reduction potentials:

$Ag^+(aq) + e^- \rightleftharpoons Ag(s)$ $\quad E^\ominus = +0.80\,V$

$Ni^{2+}(aq) + 2e^- \rightleftharpoons Ni(s)$ $\quad E^\ominus = -0.25\,V$

(a) Using the electrode potentials, write down the spontaneous processes that are expected to occur and identify the negative electrode.
(b) State which species has been reduced and which has been oxidised.
(c) Write down the conventional cell diagram for the process taking place.
(d) Calculate the overall electrode potential.

Answer

(a) Because the Ag^+/Ag electrode potential is more positive than the Ni^{2+}/Ni electrode potential, the silver process will proceed from left to right as written above. Electrons are required in this process and these are gained from the nickel metal. This forces the nickel half-reaction to move from right to left. So the spontaneous processes will be:

$Ag^+(aq) + e^- \rightarrow Ag(s)$

This is the positive electrode because electrons are being removed from it.

$Ni(s) \rightarrow Ni^{2+}(aq) + 2e^-$

This is the negative electrode because electrons are being released.

(b) Silver ions are therefore reduced to silver, and nickel is oxidised to nickel(II) ions.

(c) Figure 18.3 shows the conventional representation for the spontaneous process, with the polarity of the terminals shown.

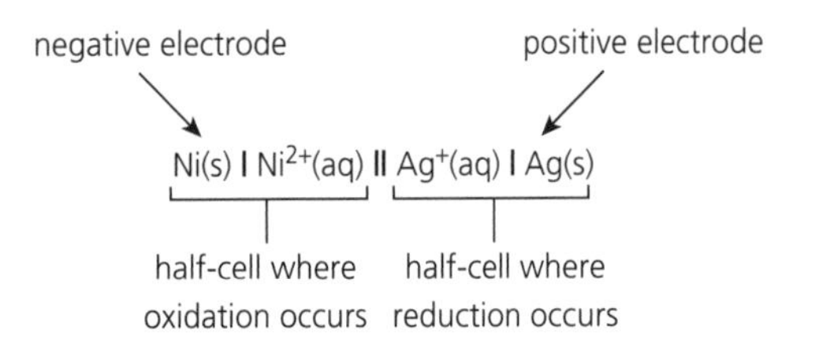

Figure 18.3 **Cell representation**

(d) The overall electrode potential, $E^\ominus_{cell}$, is calculated as follows:

$E^\ominus_{cell} = E^\ominus_{rhs} - E^\ominus_{lhs}$

$= +0.80 - (-0.25)$

$= +1.05\,V$

Exam tip

In a cell diagram, each half must balance but it is not necessary to balance the charge by writing a '2' in front of the silver ion and the silver atom on the right in Figure 18.3.

Now test yourself

TESTED

4 Given the electrochemical series below:

	$E^\ominus$/V
$F_2(g) + 2e^- \rightarrow 2F^-(aq)$	+2.87
$H_2O(aq) + 2H^+(aq) + 2e^- \rightarrow 2H_2O(l)$	+1.77
$Au^+(aq) + e^- \rightarrow Au(s)$	+1.68
$Cl_2(g) + 2e^- \rightarrow 2Cl^-(aq)$	+1.36
$O_2(g) + 4H^+(aq) \rightarrow 4e^- \rightarrow 2H_2O(l)$	+1.23
$Br_2(l) + 2e^- \rightarrow 2Br^-(aq)$	+1.09

Indicate whether each of the following pairs will result in a reaction or not.

(a) (i) $Br_2(l)$ and $Cl^-(aq)$
(ii) $H_2O_2(aq)$ in acidic solution and $Br^-(aq)$
(iii) $Au^+(aq)$ and $Br_2(l)$
(iv) $F^-(aq)$ and $Br_2(l)$
(v) $O_2(g)$ in acidic solution and $Br^-(aq)$

(b) A cell is set up involving the electrode systems Au^+/Au ($E^\ominus$ = 1.68 V) and Zn^{2+}/Zn ($E^\ominus$ = −0.76 V).
(i) Draw a labelled diagram of the cell formed when each half-cell is combined.
(ii) Calculate the overall cell potential.
(iii) Give the conventional representation of this cell.

Answers on pp. 220–221

Electrochemical cells

In an electrochemical cell, there takes place an electrochemical process from which electrical energy can be obtained and then used for doing useful work. All electrochemical cells contain two halves — oxidation takes place in one half and reduction in the other.

Many cells are **non-rechargeable** because their electrochemical process is irreversible. However, other cells use processes that are reversible and these can be recharged.

A lead storage cell is an example of a reversible process, so these cells can be recharged. The half-reactions taking place at the electrodes during discharge (the spontaneous process) are as follows:

$$PbO_2(s) + 3H^+(aq) + HSO_4^-(aq) + 2e^- \rightleftharpoons PbSO_4(s) + 2H_2O(l)$$

$$Pb(s) + HSO_4^-(aq) \rightleftharpoons PbSO_4(s) + H^+(aq) + 2e^-$$

Adding these two together gives the overall process taking place:

$$Pb(s) + PbO_2(s) + 2H_2SO_4(aq) \rightleftharpoons 2PbSO_4(s) + 2H_2O(l)$$

The forward reaction describes the spontaneous discharging process and the reverse reaction describes the charging process.

Lithium ion cells

REVISED

Lithium ion cells are used to power cameras and mobile phones. A simplified representation of a cell is shown below:

$Li \mid Li^+ \parallel Li^+, CoO_2 \mid LiCoO_2 \mid Pt$

The reagents in the cell are absorbed onto powdered graphite, which acts as a support medium. Water is not used as a solvent because it can react violently with lithium, producing hydrogen gas, and the reaction is highly exothermic too.

The support medium allows the ions to react in the absence of a solvent such as water. The half-equation for the reaction at the positive electrode (the right-hand side of the cell) can be represented as follows:

$Li^+ + CoO_2 + e^- \rightarrow Li^+[CoO_2]^-$

The oxidation state of lithium has stayed the same, at +1, so no reduction or oxidation has taken place here. However, the cobalt in CoO_2 has an oxidation state of +4, and in the cobalt species formed, $[CoO_2]^-$ it is +3. So one electron is required to enable this process to occur; this is a **reduction**, because an electron has been gained.

At the negatively charged electrode (the left-hand side of the cell), lithium atoms lose an electron to become lithium ions, and are therefore oxidised:

$Li \rightarrow Li^+ + e^-$

It can also be seen that platinum is used as an electrode on the right-hand side. Platinum is useful for this process because it is chemically unreactive and so does not become involved in the electrochemical process. It is also a very good electrical conductor.

Fuel cells

REVISED

In a typical cell, the potential difference decreases with time because the concentrations in the half-cells change as the spontaneous cell reaction takes place. This continues until the electrochemical processes are in equilibrium, and the cell can no longer be used.

In a **fuel cell**, the 'fuel' can be added constantly, so that the current is constant with time. The cell therefore does not need charging all the time.

The hydrogen–oxygen fuel cell has been shown to be very useful. In this cell, hydrogen is the fuel and it undergoes the process of oxidation at the anode. Electrons are released that then pass round the external circuit, do work, and are then used to reduce oxygen molecules. These reactions can happen either in acidic or alkaline conditions:

- in acid:
  anode: $H_2(g) \rightarrow 2H^+(aq) + 2e^-$
  cathode: $O_2(g) + 4H^+(aq) + 4e^- \rightarrow 2H_2O(l)$
- in alkali:
  anode: $2H_2(g) + 4OH^-(aq) \rightarrow 4H_2O(l) + 4e^-$
  cathode: $O_2(g) + 2H_2O(l) + 4e^- \rightarrow 4OH^-(aq)$

As usual, the overall process can be obtained by adding the half-equations for the individual electrochemical processes, so that the electrons cancel:

$2H_2(g) + O_2(g) \rightarrow 2H_2O(l)$

Exam tip

In examinations, you may be asked to add together two half-equations to form an overall equation. Alternatively you may be asked to work out a missing half-equation if you are provided with the overall equation and the other half-equation.

Advantages and disadvantages of fuel cells

Advantages

Fuel cells:

- are much more energy-efficient than conventional methods of energy provision
- are less complex than conventional gas or diesel engines
- are not subject to the high temperatures or structural weaknesses found in other engines
- will operate indefinitely as long as the fuel is available
- produce no pollutants — the only emission from a hydrogen–oxygen fuel cell is water

Disadvantages

- Hydrogen requires costly and fairly inefficient methods of extraction from existing sources. Its production could also release carbon dioxide as one of the by-products.
- Despite the fact that hydrogen is a low-density gas and any leaks can be dissipated quickly, there remains a serious risk of fire and, potentially, explosion.
- Storing hydrogen is a problem — the fuel tank would have to be large, under pressure and thermally insulated (hydrogen boils at −252°C). The gas could possibly be absorbed into a material like palladium, or adsorbed onto some metallic surfaces.

Exam practice

1 Redox reactions occur during the discharge of all electrochemical cells. Some of these cells are of commercial value.
The table below gives some redox half-equations and standard electrode potentials.

Half-equation	$E^{\ominus}$/V
$Cr^{2+}(aq) + 2e^- \rightleftharpoons Cr(s)$	−0.91
$AgCl(s) + e^- \rightleftharpoons Ag(s) + Cl^-(aq)$	+0.22
$O_2(g) + 4H^+(aq) + 4e^- \rightleftharpoons 2H_2O(l)$	+1.23
$O_3(g) + 2H^+(aq) + 2e^- \rightleftharpoons O_2(g) + H_2O(l)$	+2.07

(a) In terms of electrons, state what is meant by a 'reducing agent'. [1]
(b) Use the table above to identify the strongest reducing agent from the species listed. Explain your answer. [2]
(c) Use data from the table to explain why ozone in acidic solution reacts with chromium. Write an equation for the reaction that occurs. [3]
(d) An electrochemical cell can be constructed using a chromium electrode, and an electrode in which silver and chloride ions are in contact with silver chloride. The cell can be used to power certain electronic devices.
(i) Give the conventional representation for this cell. [1]
(ii) Calculate the e.m.f. of the cell. [1]
(iii) Write the overall equation for the reaction that occurs during discharge. [1]
(iv) Suggest *one* reason why the cell cannot be recharged electrically. [1]

2 The electrode half-equations in a lead–acid cell are given in the table below.

Half-equation	$E^\ominus$/V
$PbO_2(s) + 3H^+(aq) + HSO_4^-(aq) + 2e^- \rightleftharpoons PbSO_4(s) + 2H_2O(l)$	+1.69
$PbSO_4(s) + H^+(aq) + 2e^- \rightleftharpoons Pb(s) + HSO_4^-(aq)$	

(a) The $PbO_2/PbSO_4$ electrode is the positive terminal of the cell and the e.m.f. of the cell is 2.15 V.
Use the information to calculate the missing electrode potential indicated in the table. [1]

(b) A lead–acid cell can be recharged. Write an equation for the overall reaction that occurs when the cell is being recharged. [2]

Answers and quick quiz 18 online

ONLINE

Summary

You should now have an understanding of:

- what is meant by a 'redox' reaction
- how to use oxidation states
- how to write half-equations for redox processes and how to combine them to form an overall equation
- conventional representations of electrochemical cells
- the standard hydrogen electrode and how it is used to measure standard electrode potentials
- what standard conditions are and how important they are when measuring standard electrode potentials
- the electrochemical series and how it can be used to predict the direction of spontaneous chemical change
- some important electrochemical cells — rechargeable, non-rechargeable and fuel cells
- the hydrogen–oxygen fuel cell and the electrochemical processes involved
- the lead–acid storage cell as an example of a rechargeable cell
- the chemical processes taking place in a lithium ion cell
- the advantages and disadvantages of fuel cells

19 Acid–base equilibria

Brönsted–Lowry acids and bases

Reactions between acids and bases in solution involve the **transfer of a proton** (H^+) from one species to another — for example, in the reaction between hydroiodic acid, HI, and ethanoic acid, CH_3COOH:

$$HI(aq) + CH_3COOH(aq) \rightleftharpoons I^-(aq) + CH_3COOH_2^+(aq)$$

it can be seen that:

- HI has donated a proton to CH_3COOH, so HI is a **Brönsted–Lowry acid**
- CH_3COOH has accepted a proton from HI, so CH_3COOH is a **Brönsted–Lowry base** in this reaction

A **Brönsted–Lowry acid** is a proton donor; a **Brönsted–Lowry base** is a proton acceptor.

It is possible to pair up each species with the one it forms to make a conjugate acid–base pair. For example, HI is a Brönsted–Lowry acid in the forward direction, but I^- is a base in the reverse direction:

$$HI(aq) + CH_3COOH(aq) \rightleftharpoons I^-(aq) + CH_3COOH_2^+(aq)$$

acid 1 base 2 base 1 acid 2

Exam tip

The difference between an acid and its conjugate base (or vice versa) is simply one proton — for example HI (acid) and I^- (conjugate base) or CH_3COOH (base) and $CH_3COOH_2^+$ (conjugate acid).

Now test yourself TESTED

1 Label the conjugate acid–base pairs in the reaction between sulfuric(VI) acid and nitric(V) acid:

$$HNO_3(aq) + H_2SO_4(aq) \rightleftharpoons HSO_4^-(aq) + H_2NO_3^+(aq)$$

Answer on p. 221

Strong and weak acids and bases

Hydrochloric acid, HCl(aq), undergoes complete dissociation in aqueous solution — it is a **strong acid**. In other words, virtually all of the **molecular** hydrogen chloride, HCl, once dissolved in water, dissociates completely to form **ions**:

$$HCl(aq) + H_2O(l) \rightarrow H_3O^+(aq) + Cl^-(aq)$$

Strong acids and **strong bases** dissociate completely in solution.

Weak acids and **weak bases** dissociate partially in solution.

Strong bases like sodium hydroxide dissolve readily in water and are fully ionised, forming hydroxide ions, $OH^-(aq)$:

$$NaOH(s) + aq \rightarrow Na^+(aq) + OH^-(aq)$$

A soluble base is called an **alkali**.

Weak acids and **weak bases** interact with water molecules to form equilibrium mixtures.

The weak acid ethanoic acid, CH_3COOH, dissociates in water as follows:

$$CH_3COOH(aq) + H_2O(aq) \rightleftharpoons CH_3COO^-(aq) + H_3O^+(aq)$$

The weak base ammonia, NH_3, dissociates in water as follows:

$$NH_3(aq) + H_2O(l) \rightleftharpoons NH_4^+(aq) + OH^-(aq)$$

Now test yourself

TESTED

2 Write equations to show the dissociation of the following in aqueous solution:
(a) benzoic acid, C_6H_5COOH — a weak acid
(b) methylamine, CH_3NH_2 — a weak base

Answer on p. 221

pH

This is a measure of the 'acidity' (or 'alkalinity') of an aqueous solution. pH is defined mathematically:

$$pH = -\log_{10}[H^+(aq)]$$

Exam tip

Make sure that you quote pH to 2 decimal places in your answers.

Example 1

Calculate the pH of 0.00560 mol dm^{-3} HCl(aq).

Answer

$$HCl(aq) + H_2O(l) \rightarrow H_3O^+(aq) + Cl^-(aq) \text{ or } HCl(aq) \rightarrow H^+(aq) + Cl^-(aq)$$

So the concentration of hydrogen ions, $H^+(aq)$, will be 0.00560 mol dm^{-3}.

$$pH = -\log_{10}[H^+(aq)]$$

$$= -\log_{10}(0.00560)$$

$$= -(-2.25) = 2.25$$

Example 2

Calculate the pH of 0.100 mol dm^{-3} dilute sulfuric(VI) acid.

Answer

$$H_2SO_4(aq) \rightarrow 2H^+(aq) + SO_4^{2-}(aq)$$

So the concentration of hydrogen ions, $H^+(aq)$, will be (0.100 × 2) mol dm^{-3}.

$$pH = -\log_{10}[H^+(aq)]$$

$$= -\log_{10}(0.200)$$

$$= 0.70$$

Calculating hydrogen ion concentration given the pH

REVISED

Because $pH = -\log_{10}[H^+(aq)]$, rearranging this to make $[H^+(aq)]$ the subject gives:

$$[H^+(aq)] = 10^{-pH}$$

Example

Calculate the concentration of hydrogen ions in a solution that has a pH of 13.20.

Answer

$[H^+(aq)] = 10^{-13.20}$

$= 6.31 \times 10^{-14}\,mol\,dm^{-3}$

The ionic product of water, K_w

Water dissociates very slightly according to the equilibrium:

$$H_2O(l) \rightleftharpoons H^+(aq) + OH^-(aq)$$

An equilibrium constant (similar to K_c) can be written for this dissociation. It has the symbol K_w, and is called the **ionic product of water**:

$$K_w = [H^+(aq)][OH^-(aq)]$$

The value of K_w is $1.00 \times 10^{-14}\,mol^2\,dm^{-6}$ at 298 K.

The value of K_w depends, like all equilibrium constants, on temperature. As the temperature increases, so does K_w. This indicates that the dissociation of water is an endothermic process. As the temperature increases, the concentrations of $H_3O^+(aq)$ and $OH^-(aq)$ also increase.

Calculating the pH of a solution of a strong base

REVISED

The pH of a solution of a strong base can be calculated using the ionic product of water, K_w.

Example 1

Calculate the pH of $9.40 \times 10^{-3}\,mol\,dm^{-3}$ sodium hydroxide solution, NaOH(aq), at 298 K. The value of K_w is $1.00 \times 10^{-14}\,mol^2\,dm^{-6}$ at 298 K.

Answer

$[OH^-(aq)] = 9.40 \times 10^{-3}\,mol\,dm^{-3}$

$K_w = [H^+(aq)][OH^-(aq)]$

$1.00 \times 10^{-14} = [H^+(aq)] \times (9.40 \times 10^{-3})$

$[H^+(aq)] = 1.064 \times 10^{-12}\,mol\,dm^{-3}$

$pH = -\log_{10}[H^+(aq)]$

$= -\log_{10}(1.064 \times 10^{-12})$

$= 11.97$

Example 2

Calculate the pH of the mixture formed when 20.0 cm^3 of 0.200 mol dm^{-3} H_2SO_4 is added to 40.0 cm^3 of 0.250 mol dm^{-3} NaOH.

Answer

$$\text{moles of } H_2SO_4 = \frac{20.0}{1000} \times 0.200$$
$$= 4.00 \times 10^{-3}\,\text{mol}$$

$$\text{moles of NaOH} = \frac{40.0}{1000} \times 0.250$$
$$= 1.00 \times 10^{-2}\,\text{mol}$$

The equation for the reaction is:

$$2NaOH + H_2SO_4 \rightarrow Na_2SO_4 + 2H_2O$$

4.00×10^{-3} moles of H_2SO_4 will react exactly with 8.00×10^{-3} moles of NaOH (using the 1 : 2 ratio from the equation).

However, 1.00×10^{-2} moles of NaOH are present, so $(1.00 \times 10^{-2} - 8.00 \times 10^{-3})$ mol of NaOH will remain — that is 2.00×10^{-3} mol. This is present in a total volume of 60.0 cm^3.

$$[OH^-] = \frac{1000}{60.0} \times 2.00 \times 10^{-3}$$
$$= 0.0333\,\text{mol dm}^{-3}$$

But:

$$K_w = [H^+][OH^-] = 1.00 \times 10^{-14}$$

So:

$$[H^+] = \frac{K_w}{[OH^-]} = \frac{1.00 \times 10^{-14}}{0.0333}$$
$$= 3.00 \times 10^{-13}\,\text{mol dm}^{-3}$$

$$pH = -\log_{10}[H^+(aq)]$$
$$= -\log_{10}(3.00 \times 10^{-13})$$
$$= 12.52$$

Now test yourself

TESTED

3. Calculate the pH of the following solutions of strong acids and bases at 298 K.
   Assume the value of K_w is 1.00×10^{-14} mol^2 dm^{-6}
   (a) 6.60×10^{-2} mol dm^{-3} HNO_3(aq)
   (b) 5.67×10^{-4} mol dm^{-3} NaOH(aq)
   (c) 0.0500 mol dm^{-3} KOH(aq)
   (d) 0.0950 mol dm^{-3} H_2SO_4(aq).
4. A sample of rainwater is found to have a pH of 6.45. What is the concentration of hydrogen ions and the concentration of hydroxide ions in the rainwater at 298 K?

Answers on p. 221

The acid dissociation constant, K_a

A weak acid, HA, dissociates in water according to the equilibrium:

$$HA(aq) + H_2O(l) \rightleftharpoons H_3O^+(aq) + A^-(aq)$$

or:

$$HA(aq) \rightleftharpoons H^+(aq) + A^-(aq)$$

An equilibrium constant, K_a, can be written as:

$$K_a = \frac{[H_3O^+(aq)][A^-(aq)]}{[HA]}$$

or:

$$\frac{[H^+][A^-]}{[HA]}$$

K_a has units of mol dm^{-3}.

It is also possible to convert K_a values into pK_a values:

$$pK_a = -\log_{10}K_a$$

$$K_a = 10^{-pK_a}$$

As acids increase in strength, the values of K_a increase and the values of pK_a decrease (or stronger acids have larger K_a values and smaller pK_a values).

The pK_a values in Table 19.1 show acids that decrease in strength from top to bottom.

Table 19.1 Typical pK_a values

Hydrofluoric acid, HF	3.20
Benzoic acid, C_6H_5COOH	4.20
Ethanoic acid, CH_3COOH	4.76
Hydrogen cyanide, HCN	9.40

Typical mistake

Many students include water in an expression for K_a. Water is never put into a K_a expression because its concentration is taken as constant.

Typical mistake

Many students state wrongly that acidic strength increases as pK_a increases. Remember that acidic strength increases as K_a increases, but as pK_a decreases.

Calculating the pH of a solution of a weak acid

REVISED

The pH of a solution of a weak acid is calculated in a similar way to that for a solution of a strong acid — with an adjustment for the fact that incomplete dissociation occurs.

Example

Calculate the pH of a solution of butanoic acid, $CH_3(CH_2)_2COOH$, of concentration 0.150 mol dm^{-3}. K_a for butanoic acid is 1.514×10^{-5} mol dm^{-3} at 298 K.

Answer

$$CH_3(CH_2)_2COOH(aq) \rightleftharpoons H_3O^+(aq) + CH_3(CH_2)_2COO^-(aq)$$

At start/mol: 0.150 0 0

Let x be the number of moles of acid that dissociate in 1.0 dm^3 of solution:

At equilibrium/mol: $0.150 - x$ x x

Because the acid is weak, we can make the approximation that 0.150 is much larger than x, and can ignore x in the left-hand term.

Concentration at equilibrium/mol dm^{-3}:

$$\frac{0.150}{1} \qquad \frac{x}{1} \qquad \frac{x}{1}$$

$$K_a = \frac{[CH_3(CH_2)_2COO^-(aq)][H^+(aq)]}{[CH_3(CH_2)_2COOH(aq)]}$$

Substituting the values we know:

$$1.514 \times 10^{-5} = x \times \frac{x}{0.150}$$

$$x^2 = 1.514 \times 10^{-5} \times 0.150$$

$$x = \sqrt{(1.514 \times 10^{-5} \times 0.150)}$$

$$= 1.507 \times 10^{-3}\,\text{mol dm}^{-3}$$

This is equal to the hydrogen ion concentration, $[H_3O^+(aq)]$, so:

$$pH = -\log_{10}[H^+(aq)]$$

$$= -\log_{10}(1.507 \times 10^{-3})$$

$$= 2.82$$

Exam tip

It looks a long calculation, but with practice you will realise that these are not that difficult.

It is acceptable to avoid some of the algebra in the method above and use:

$$K_a = \frac{[H^+(aq)][A^-(aq)]}{[HA(aq)]}$$

For a solution of weak acid $[H^+] = [A^-]$, so:

$$K_a = \frac{[H^+]^2}{[HA]}$$

Therefore:

$$[H^+]^2 = K_a \times [HA]$$

$$[H^+(aq)] = \sqrt{(K_a \times [HA(aq)])}$$

Now test yourself

TESTED

5 (a) Write an equation to show the partial dissociation of ethanoic acid, CH_3COOH, in water.

(b) Write an expression for the acid dissociation constant for ethanoic acid.

(c) Calculate the pH of a 0.100 mol dm^{-3} aqueous solution of ethanoic acid.
(K_a for ethanoic acid is 1.74×10^{-5} mol dm^{-3})

Answer on p. 221

Titration

A titration curve (Figure 19.1) is produced when a base is titrated against an acid and the pH is monitored as the volume of added base increases.

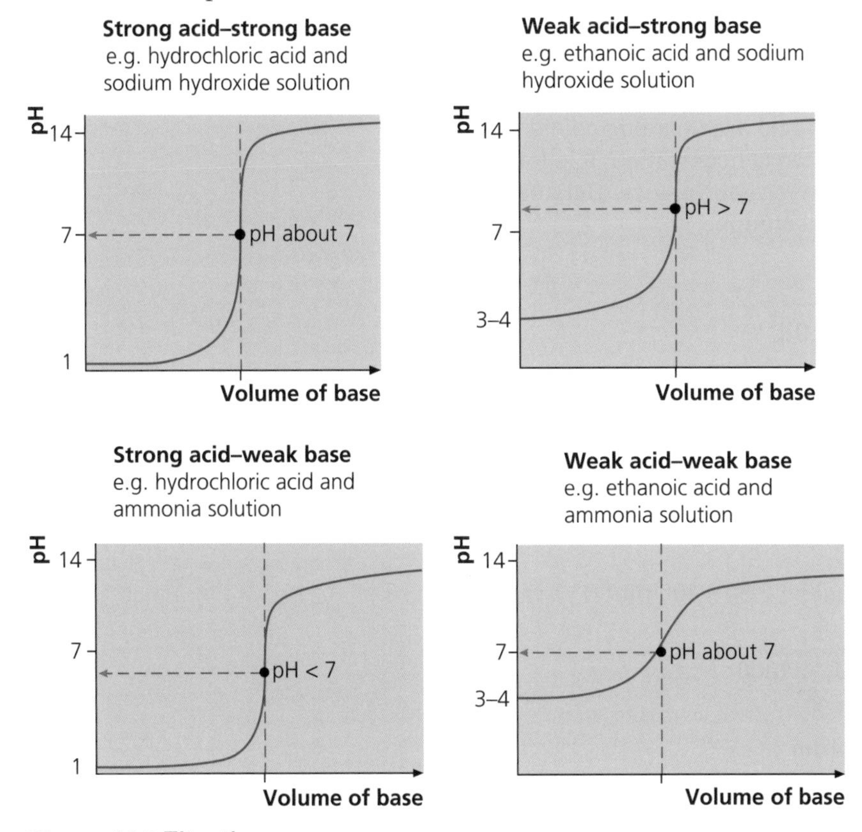

Figure 19.1 Titration curves

The shapes of these curves depend on many factors:

- The starting and final pH values depend on the strength of the acid and/or base used. For example, if hydrochloric acid (a strong acid) is used then the starting pH will be approximately equal to 1; if ammonia solution (a weak base) is used then a final pH of 9–10 will result.
- The steep part of the curve, where there is a rapid change of pH, shows the volume of base needed for **equivalence** with the amount of acid used.
- The equivalence pH depends on the strengths of the acid and base used and the nature of the salt formed at the end-point — Table 19.2 shows some typical cases.

Table 19.2 Typical titration pairs at equivalence points

Type	pH	Example of salt formed
Strong acid–strong base	About 7	Sodium chloride
Strong acid–weak base	Lower than 7	Ammonium chloride
Weak acid–strong base	Higher than 7	Sodium ethanoate
Weak acid–weak base	About 7	Ammonium ethanoate

Titration calculations

REVISED

Questions may be set that involve simple titration data.

Example 1

Involving an unknown concentration

25.0 cm^3 of 0.100 mol dm^{-3} hydrochloric acid are titrated against sodium hydroxide solution of an unknown concentration. It is found that 23.50 cm^3 of the alkali are required for equivalence. Calculate the concentration of the sodium hydroxide solution.

Answer

$$NaOH(aq) + HCl(aq) \rightarrow NaCl(aq) + H_2O(l)$$

$$\text{moles of HCl(aq)} = \frac{\text{volume (cm}^3\text{)}}{1000\,\text{cm}^3} \times \text{concentration (mol dm}^{-3}\text{)}$$

$$= \frac{25.0}{1000} \times 0.100$$

$$= 2.50 \times 10^{-3}\,\text{mol}$$

So the number of moles of NaOH must be 2.50×10^{-3} mol (1 : 1 ratio in the equation).

$$\text{concentration of NaOH} = \frac{2.50 \times 10^{-3} \times 1000}{23.50}$$

$$= 0.106\,\text{mol dm}^{-3}$$

Example 2

Involving an unknown volume

25.0 cm^3 of 0.250 mol dm^{-3} sulfuric(VI) acid are titrated against potassium hydroxide solution of concentration 0.825 mol dm^{-3}. What volume of the potassium hydroxide solution is required for equivalence?

Answer

$$2KOH(aq) + H_2SO_4(aq) \rightarrow K_2SO_4(aq) + 2H_2O(l)$$

$$\text{moles of H}_2\text{SO}_4\text{(aq)} = \frac{\text{volume (cm}^3\text{)}}{1000\,\text{cm}^3} \times \text{concentration (mol dm}^{-3}\text{)}$$

$$= \frac{25.0}{1000} \times 0.250$$

$$= 6.25 \times 10^{-3}\,\text{mol}$$

So the amount of KOH must be $2 \times 6.25 \times 10^{-3}$ mol = 0.0125 mol (2 : 1 ratio in the equation).

$$\text{volume of KOH} = \frac{0.0125 \times 1000}{0.825}$$

$$= 15.15\,\text{cm}^3$$

Indicators

REVISED

Indicators are weak acids or weak bases that are able to give a measure of the pH of a solution by their colour. There are many different indicators, which change from one colour to another at different pH values — these are called pK_{in} values.

Figure 19.2 shows the colours of some well-known indicators, together with the pH values at which they change colour.

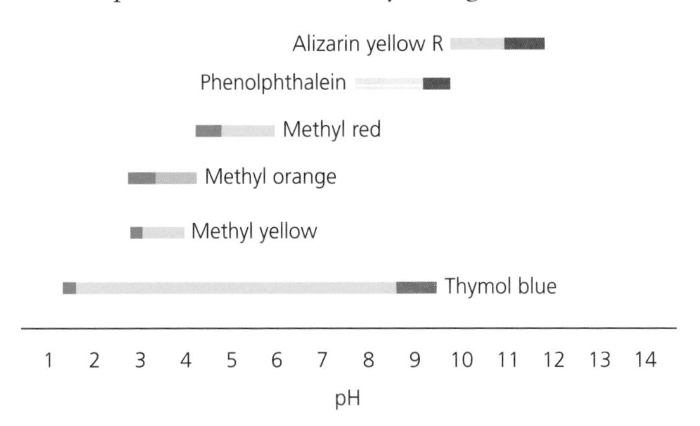

Figure 19.2 Some indicators and their colour change ranges

Choice of indicator

REVISED

The choice of just which indicator to use in a particular titration depends on the pH at the equivalence point in the titration — the pH at the point at which the acid and base have reacted with each other exactly.

We choose an indicator that changes colour at the pH equal to, or very close to, the pH at the equivalence point of the acid–base reaction. This is always on the 'steep' part of the titration curve. This means that the colour change occurs with the addition of one drop of base and that the equivalence point is sharp.

In the titration shown in Figure 19.3, between a strong acid and a weak base, the pH at equivalence is about 5. So we need to choose an indicator that changes colour at about this pH — from Figure 19.2 methyl orange looks a good choice. If we chose phenolphthalein, which has a pK_{in} value indicating a colour change at 8–10, this would result in the indicator changing over the addition of a large volume of base, with no sharp end-point.

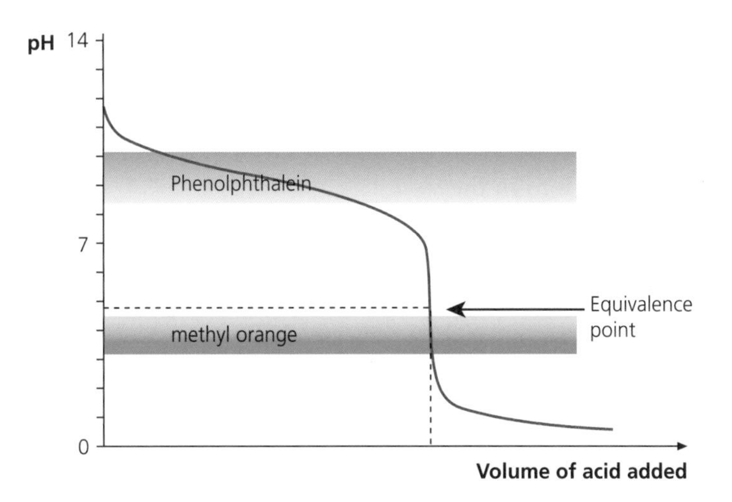

Figure 19.3 A typical titration curve for the neutralisation of a weak base by a strong acid

Exam tip

Choose an indicator so that its pK_{in} value is never more than 1 pH unit from the equivalence pH. So, if the pH at equivalence is 9.20, the pK_{in} of the indicator should be between 8.20 and 10.20.

Buffer solutions

Buffer solutions exist as one of two main types according to their components.

- **Acid buffers:** a solution containing a mixture of a weak acid and its conjugate base. The pH of the buffer solution is lower than 7. Example: ethanoic acid, $CH_3COOH(aq)$, and sodium ethanoate, $CH_3COO^-Na^+$.
- **Basic buffers**: a solution containing a mixture of a weak base and its conjugate acid. The pH of the buffer solution is higher than 7. Example: ammonia solution, $NH_3(aq)$, and ammonium chloride, NH_4Cl.

The pH at which a buffer operates most effectively depends on the pK_a of the acidic component present.

For a buffer to operate effectively, it is crucial that the mixture contains large amounts of both the base and the acid.

Buffer solutions are defined as solutions that resist a change in pH on adding small amounts of either acid or base (or by diluting the solution by adding water).

Exam tip

Remember that any carboxylic acid and one of its corresponding salts can be used (when combined) to make a buffer.

How does a buffer work?

REVISED

Consider the effect of adding acid and base to an ethanoic acid–sodium ethanoate buffer system (an acid buffer). The equilibrium operating is:

$$CH_3COOH(aq) + H_2O(l) \rightleftharpoons CH_3COO^-(aq) + H_3O^+(aq)$$

Exam tip

Always write an equation for the dissociation of the *acid* — a carboxylic acid or ammonium ion — and then use that equilibrium to explain how the buffer works

On adding a base

Added hydroxide ions react with the hydrogen ions, $H_3O^+(aq)$, in the above equilibrium:

$$H_3O^+(aq) + OH^-(aq) \rightarrow 2H_2O(l)$$

This reduces the hydrogen ion concentration in the equilibrium mixture. The equilibrium consequently shifts to the right-hand side by dissociating some of its molecular ethanoic acid to produce more hydrogen ions (Le Chatelier principle):

$$CH_3COOH(aq) + H_2O(l) \rightarrow CH_3COO^-(aq) + H_3O^+(aq)$$

The final pH in the new equilibrium mixture is similar to its starting pH and the value for K_a in the equilibrium mixture is preserved (remember that changes in concentration do not affect K_a or any other equilibrium constant).

On adding an acid

$$CH_3COOH(aq) + H_2O(l) \rightleftharpoons CH_3COO^-(aq) + H_3O^+(aq)$$

Adding an acid increases the concentration of the hydrogen ions in the equilibrium above. The equilibrium shifts to the left-hand side, using up ethanoate ions, CH_3COO^-, to remove the excess hydrogen ions:

$$CH_3COO^-(aq) + H_3O^+(aq) \rightarrow CH_3COOH(aq) + H_2O(l)$$

The concentration of hydrogen ions in the new equilibrium mixture is similar to that at the start and the pH remains about the same.

Another commonly used buffer is the ammonia–ammonium system. The equilibrium operating here is:

$$NH_4^+(aq) + H_2O(l) \rightleftharpoons NH_3(aq) + H_3O^+(aq)$$

On adding acid, the reaction shifts to the left-hand side; on adding alkali, the reaction shifts to the right-hand side. The explanation is the same as for the ethanoic acid–sodium ethanoate system.

Calculations involving buffer solutions

REVISED

Consider the general dissociation of a weak acid, HA:

$$HA(aq) \rightleftharpoons A^-(aq) + H^+(aq)$$

$$K_a = \frac{[H^+(aq)][A^-(aq)]}{[HA(aq)]}$$

A question may require you to calculate the pH of a buffer solution in which the concentrations of both components are given, that is, $[A^-]$ and [HA], together with pK_a for the acid present.

Example

Calculate the pH of a buffer solution that contains a mixture of 0.0500 mol dm^{-3} HCOOH(aq) and 0.0800 mol dm^{-3} HCOONa(aq).The pK_a of methanoic acid is 3.75.

Answer

$$HCOOH(l) \rightleftharpoons HCOO^-(aq) + H^+(aq)$$

$$K_a = \frac{[HCOO^-(aq)][H^+(aq)]}{[HCOOH(aq)]}$$

pK_a for the acid is 3.75, so:

$$K_a = 10^{-3.75}$$

$$= 1.78 \times 10^{-4}\ mol\,dm^{-3}$$

Substituting into the expression for K_a gives:

$$1.78 \times 10^{-4} = \frac{0.0800 \times [H^+(aq)]}{0.0500}$$

$$[H^+(aq)] = \frac{(1.78 \times 10^{-4} \times 0.0500)}{0.0800}$$

$$= 1.11 \times 10^{-4}\ mol\,dm^{-3}$$

$$pH = -log_{10}[H^+(aq)]$$

$$= -log_{10}(1.11 \times 10^{-4})$$

$$= 3.95$$

Now test yourself

TESTED

6 A buffer solution is required with a pH of 4.50. A solution containing benzoic acid (C_6H_5COOH) and sodium benzoate (C_6H_5COONa) is used to prepare this buffer. The pK_a value for benzoic acid is 4.20.
   (a) Write an equation to show benzoic acid dissociating in water, including state symbols.
   (b) Calculate the ratio of the benzoic acid and sodium benzoate concentrations that would give a pH of 4.50. Give your answer to 3 significant figures.
   (c) What would happen to the value of benzoic acid's pK_a when the following changes are made?
      (i) A solution of 1.00 mol dm^{-3} sodium hydroxide is added to the buffer solution
      (ii) The temperature is increased?

Answers on p. 221

Exam practice

1 This question is about the pH of some solutions containing potassium hydroxide or methanoic acid, or both. Give all pH values to 2 decimal places.
   (a) (i) Write an expression for pH. [1]
      (ii) Write an expression for the ionic product of water. [1]
      (iii) At 10°C, a 0.132 mol dm^{-3} solution of potassium hydroxide has a pH of 13.22. Calculate the value of K_w at 10°C. [2]
   (b) (i) Write an expression for K_a for methanoic acid, HCOOH(aq). [1]
      (ii) Calculate the pH of a solution of methanoic acid of concentration 0.200 mol dm^{-3}. pK_a for methanoic acid is 3.75. [3]
   (c) (i) What is meant by a 'buffer solution'? [1]
      (ii) Calculate the pH of the solution formed when 20.0 cm^3 of 0.200 mol dm^{-3} methanoic acid is added to 10.0 cm^3 of 0.132 mol dm^{-3} potassium hydroxide solution. [4]

2 A titration is carried out in which 25.0 cm^3 of benzoic acid of concentration 0.100 mol dm^{-3} is titrated against sodium hydroxide solution of an unknown concentration. It is found that 15.50 cm^3 of sodium hydroxide is required for complete reaction. (pK_a for benzoic acid is 4.20)
   (a) (i) Write an equation for the reaction taking place. [1]
      (ii) Calculate the concentration of the sodium hydroxide solution used. [2]
      (iii) Sketch a titration curve for the reaction taking place. Label your axes carefully, indicating the pH at which equivalence takes place. [2]
      (iv) Give the name of an indicator that could be used in this titration. Select from the list in Figure 19.2 and explain your choice. [2]
   (b) In another titration, 25.0 cm^3 of ethanoic acid (pK_a = 4.76) of concentration 0.100 mol dm^{-3} was used instead of 25.0 cm^3 of benzoic acid. State how the volume of sodium hydroxide solution (of the same concentration) required for neutralisation would change, if at all. [2]

Answers and quick quiz 19 online

ONLINE

Summary

You should now have an understanding of:

- Brönsted–Lowry acids and bases
- pH
- how to calculate the pH of solutions of strong acids and strong bases
- the ionic product of water, K_w
- weak acids and weak bases
- the acid dissociation constant, K_a
- how to calculate the pH of a solution of a weak acid
- pH titration curves
- indicators
- buffers and how to calculate the pH of an acid buffer solution

20 Periodicity

Periodicity is the regular and repeating pattern of physical and chemical properties when elements are arranged in order of atomic number in the periodic table. The outer electronic configuration of atoms is a periodic function, so we should expect other properties to change accordingly.

Periodicity can be demonstrated using the reactions and properties of the Period 3 elements:

Na Mg Al Si P S Cl Ar

Reactions of Period 3 elements

Reactions with water

REVISED

The notable reactions of the elements with water apply largely to sodium and magnesium.

Sodium

Sodium reacts violently with water to form hydrogen gas and a solution of a **strong alkali** (sodium hydroxide) of pH 13–14. In the reaction, sodium melts, fizzes and moves around the surface of the water, forming a colourless solution and colourless gas:

$$2Na(s) + 2H_2O(l) \rightarrow 2NaOH(aq) + H_2(g)$$

In the reaction, sodium atoms lose their outer electrons and are **oxidised** to form Na^+ ions:

$$Na(s) \rightarrow Na^+(aq) + e^-$$

Water molecules gain electrons and are reduced to form hydrogen and hydroxide ions:

$$2H_2O(l) + 2e^- \rightarrow 2OH^-(aq) + H_2(g)$$

Magnesium

Magnesium reacts only very slowly with cold water to form magnesium hydroxide — a slightly soluble white solid — and a colourless gas:

$$Mg(s) + 2H_2O(l) \rightarrow Mg(OH)_2(aq) + H_2(g)$$

It reacts more rapidly with steam to form magnesium oxide, a white solid, and hydrogen gas:

$$Mg(s) + H_2O(g) \rightarrow MgO(s) + H_2(g)$$

The oxides of Period 3 elements

The study of the physical and chemical properties of the oxides of the Period 3 elements reveals some important trends in behaviour, which can be extended to other parts of the periodic table.

Formation of the oxides

REVISED

An element can be reacted with oxygen by simply heating it in air until it ignites. The burning sample is then put into a gas jar containing pure oxygen. Alternatively, a stream of pure oxygen can be passed over a heated sample of the element in a glass tube.

A summary of the reactions of the Period 3 elements with oxygen is shown in Table 20.1.

Table 20.1 Reactions of Period 3 elements with oxygen

Element	Na	Mg	Al	Si	P	S
Formula and state of oxide	$Na_2O(s)$	$MgO(s)$	$Al_2O_3(s)$	$SiO_2(s)$	$P_4O_{10}(s)$	$SO_2(g)$ $SO_3(g)$
Equation	$4Na(s) + O_2(g) \rightarrow 2Na_2O(s)$	$2Mg(s) + O_2(g) \rightarrow 2MgO(s)$	$4Al(s) + 3O_2(g) \rightarrow 2Al_2O_3(s)$	$Si(s) + O_2(g) \rightarrow SiO_2(s)$	$P_4(s) + 5O_2(g) \rightarrow P_4O_{10}(s)$	$S(s) + O_2(g) \rightarrow SO_2(g)$
Observations	Melts and then ignites with an orange flame; a white solid is formed	A bright, white flame is formed, and a white solid remains	Fine powder reacts to form white 'sparks'; a solid is formed as a fine white powder	No observable reaction under normal conditions	A bright white/yellow flame, together with white fumes that collect to form a white solid	The yellow solid melts, then slowly disappears as it burns with a blue flame — fumes are formed
Bonding of oxide	Ionic	Ionic	Ionic with covalent character	Covalent	Covalent	Covalent
Structure of oxide	Giant ionic	Giant ionic	Giant ionic with covalent character	Giant covalent	Simple covalent	Simple covalent

Now test yourself

TESTED

1 Write equations to show the oxides formed when the following elements are heated in excess oxygen. (You may want to cover up Table 20.1 to test your memory.)
 (a) aluminium
 (b) sulfur
 (c) sodium
2 Write an equation to show what is likely to happen when caesium, Cs, a highly reactive Group 1 metal, is added to water. Comment on the pH of the solution formed.

Answers on p. 221

Reactions of Period 3 oxides with water

REVISED

Table 20.2 summarises the important observations made when the oxides react with water.

Table 20.2 Period 3 oxides with water

Formula of oxide	$Na_2O(s)$	$MgO(s)$	$Al_2O_3(s)$	$SiO_2(s)$	$P_4O_{10}(s)$	$SO_2(g)$ $SO_3(g)$
Reaction of oxide with water	White solid reacts rapidly forming a colourless solution, and releasing heat	A slow reaction in which some white solid dissolves	No observable change takes place	No observable change takes place	A vigorous reaction takes place; the white solid rapidly reacts to form a colourless solution	Both gases react with water to form colourless solutions
pH of solution formed	13–14	8–10	7	7	2–4	$SO_2(g)$: 3–5 $SO_3(g)$: 1–3
Formula of products	$NaOH(aq)$	$Mg(OH)_2(aq)$	–	–	$H_3PO_4(aq)$	$H_2SO_3(aq)$ $H_2SO_4(aq)$

Structure and bonding of Period 3 oxides

REVISED

The bonding character of the oxides changes from ionic to covalent on moving from left to right in the periodic table (Figure 20.1).

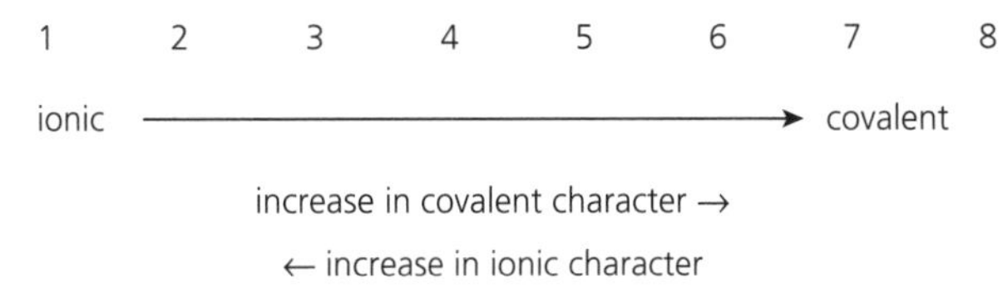

Figure 20.1 Changes in bonding character

As the bonding character changes, the physical and chemical properties also change.

Physical properties of the oxides

- **Ionic oxides**, like Na_2O, MgO and Al_2O_3, all have **giant ionic structures** similar to that shown in Figure 20.2. They all have high melting and boiling points because of the strong electrostatic forces acting between the oppositely charged ions. When solid, the oxides are poor electrical conductors (insulators) because the ions are not mobile, but when molten they conduct electricity well because the ions are mobile.

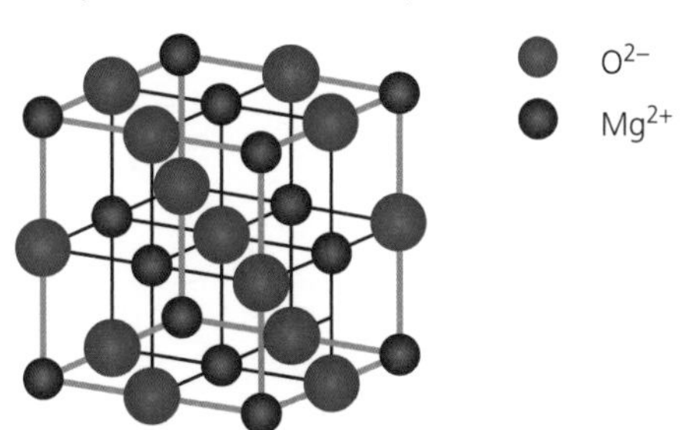

Figure 20.2 The giant ionic structure of MgO

- **Covalent oxides** like SiO_2 have a **giant covalent structure**, as shown in Figure 20.3. The silicon atoms and oxygen atoms are bonded throughout by strong Si–O bonds. These are difficult to break and hence these oxides have high melting and boiling points.
- **Covalent oxides**, like P_4O_{10}, SO_2 and SO_3, all have **simple covalent structures** and therefore exist as simple molecules weakly bonded by van der Waals forces and dipole–dipole interactions. Their melting and boiling points are low because these intermolecular forces are easily overcome. Electrical conductivity is poor because all molecules have a zero charge overall.

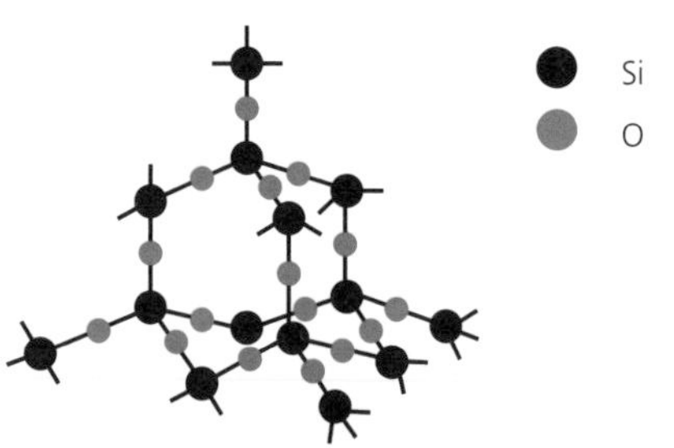

Figure 20.3 The giant structure of silicon(IV) oxide

Chemical properties of Period 3 oxides

The nature of the reaction of an oxide with water gives information about the bonding in the compound.

Generally, metal oxides (ionic) react with water to form hydroxide solutions with a high pH value. Non-metal oxides (covalent) form acidic solutions with low pH values. Both types of reaction involve a process called **hydrolysis**.

Hydrolysis is the splitting up of a substance by water. In this process, water molecules are chemically changed.

- Sodium oxide, $Na_2O(s)$, forms a solution of the strong alkali sodium hydroxide:

  $$Na_2O(s) + H_2O(l) \rightarrow 2NaOH(aq)$$

  or

  $$O^{2-}(s) + H_2O(l) \rightarrow 2OH^-(aq)$$

- Magnesium oxide, MgO(s), reacts only slightly with water to form a dilute alkaline solution of magnesium hydroxide. The product is only slightly soluble in water and so forms a low concentration of hydroxide ions:

  $$MgO(s) + H_2O(l) \rightleftharpoons Mg(OH)_2(aq)$$

- Aluminium oxide is virtually insoluble in water and does not react with it.

From silicon onwards, the oxides are covalent and exist either as giant molecules (e.g. SiO_2) that are not hydrolysed by water, or as simple molecules (e.g. SO_2, P_4O_{10}) that are hydrolysed to form acidic solutions.

- Phosphorus(V) oxide, $P_4O_{10}(s)$, is hydrolysed by water to form a solution of phosphoric(V) acid:

  $$P_4O_{10}(s) + 6H_2O(l) \rightarrow 4H_3PO_4(aq)$$

  Phosphoric(V) acid is a relatively strong acid, and when it dissociates it loses each proton in three separate stages to eventually form the phosphate(V) ion, PO_4^{3-}. The structures of a phosphoric(V) acid molecule and the phosphate(V) ion are shown in Figure 20.4.

(a)

OH
P
O OH
OH

(b)

O
P
$^-$O O$^-$
O$^-$

Figure 20.4 The structures of a phosphoric(V) acid molecule (a) and the phosphate(V) ion (b)

- Sulfur(IV) oxide, $SO_2(g)$, is hydrolysed to form sulfuric(IV) acid, $H_2SO_3(aq)$ — a weak acid:

  $SO_2(g) + H_2O(l) \rightarrow H_2SO_3(aq)$

- Sulfur(VI) oxide, $SO_3(g)$, is rapidly hydrolysed to form sulfuric(VI) acid, $H_2SO_4(aq)$ — a strong acid:

  $SO_3(g) + H_2O(l) \rightarrow H_2SO_4(aq)$

  Sulfuric(VI) acid, H_2SO_4, is a strong acid because both protons dissociate from the molecule relatively easily to form the hydrogen sulfate(V) ion, HSO_4^-, followed by the sulfate(VI) ion, SO_4^{2-}. The structures of a sulfuric(VI) acid molecule and a sulfate(VI) ion are shown in Figure 20.5.

(a) O=S(=O)(OH)OH

(b) O=S(=O)(O⁻)O⁻

Figure 20.5 The structures of a sulfuric(VI) acid molecule (a) and a sulfate(VI) ion (b)

Now test yourself

TESTED

3 Write equations to show the reactions taking place when the following oxides are added to water:
  (a) sulfur(IV) oxide
  (b) sodium oxide
4 Write an equation for the reaction between the two solutions formed in question 3.

Answers on p. 221

Acid–base properties of Period 3 oxides

Because metal oxides are bases, they will react with acids to form a salt and water. Conversely, non-metal oxides are acids and will react with bases to form a salt and water.

Table 20.3 summarises the reactions. Note that aluminium oxide reacts with both acids and bases to form salts. Oxides of this type are called **amphoteric** oxides.

Table 20.3 Reactions of Period 3 oxides with acids and alkalis

Oxide	Reaction with dilute HCl(aq)	Reaction with dilute NaOH(aq)
$Na_2O(s)$	$Na_2O(s) + 2HCl(aq) \rightarrow 2NaCl(aq) + H_2O(l)$	No reaction
$MgO(s)$	$MgO(s) + 2HCl(aq) \rightarrow MgCl_2(aq) + H_2O(l)$	No reaction
$Al_2O_3(s)$	$Al_2O_3(s) + 6HCl(aq) \rightarrow 2AlCl_3(aq) + 3H_2O(l)$	$Al_2O_3(s) + 2NaOH(aq) + 3H_2O(l) \rightarrow 2NaAl(OH)_4(aq)$
$SiO_2(s)$	No reaction	$SiO_2(s) + 2NaOH(aq) \rightarrow Na_2SiO_3(aq) + H_2O(l)$
$P_4O_{10}(s)$	No reaction	$P_4O_{10}(s) + 12NaOH(aq) \rightarrow 4Na_3PO_4(aq) + 6H_2O(l)$
$SO_2(g)$	No reaction	$SO_2(g) + 2NaOH(aq) \rightarrow Na_2SO_3(aq) + H_2O(l)$
$SO_3(s)$	No reaction	$SO_3(g) + 2NaOH(aq) \rightarrow Na_2SO_4(aq) + H_2O(l)$

Exam practice

1 (a) Describe what you would observe when magnesium burns in oxygen. Write an equation for the reaction that occurs. State the type of bonding in the oxide formed. [4]

(b) Describe what you would observe when sulfur burns in oxygen. Write an equation for the reaction that occurs. State the type of bonding in the oxide formed. [4]

(c) The substance formed in part (a) of this question is added to water, forming solution A. In a separate experiment, sulfur(VI) oxide is added to water to form solution B. Solution B is then added to solution A.

Write equations for these reactions:

(i) the oxide from part (a) reacting with water to form solution A [1]

(ii) sulfur(VI) oxide reacting with water to form solution B [1]

(iii) solution A reacting with solution B. [1]

(d) Outline an experiment to show that aluminium oxide is composed of ions. [2]

(e) Write equations to show how aluminium oxide reacts with:

(i) sulfuric(VI) acid [1]

(ii) aqueous potassium hydroxide [1]

Answers and quick quiz 20 online

ONLINE

Summary

You should now have an understanding of:

- the reactions of magnesium and sodium with water
- the reactions of Period 3 elements with oxygen and of the oxides that are formed
- how the physical properties of the oxides depend on the bonding and structure of the compounds
- the reactions of Period 3 oxides with water and the meaning of the term hydrolysis
- the fact that some metal oxides form alkaline solutions, whereas some non-metal oxides form acidic solutions
- the reactions of metal oxides with acids to form salts
- the reactions of non-metal oxides with bases to form salts
- the fact that there are oxides that are amphoteric

21 Transition metals

Electronic configurations

The first series of **transition metals** is:

$_{21}Sc$ $_{22}Ti$ $_{23}V$ $_{24}Cr$ $_{25}Mn$ $_{26}Fe$ $_{27}Co$ $_{28}Ni$ $_{29}Cu$ $_{30}Zn$

Although all transition metals are in the d-block of the periodic table, not all d-block elements are transition metals — for example, zinc is not.

A **transition metal** is defined as a metal that, in at least one of its stable ions, has a partly filled d sublevel.

Exam tip

Zinc is not regarded as a transition metal because it forms only one stable ion, Zn^{2+}, in which it has a completely filled 3d sublevel.

Electronic configurations of atoms

REVISED

When writing the electronic configurations of these elements, the order in which the sublevels fill up is important:

1s, 2s, 2p, 3s, 3p, 4s, 3d

Notice that the 4s sublevel fills before the 3d sublevel.

Example

Write the electronic configurations of titanium, manganese and nickel.

Answer

Filling up the sublevels in order of energy, lowest first, gives:

$_{22}Ti$ $1s^2, 2s^2, 2p^6, 3s^2, 3p^6, 4s^2, 3d^2$

$_{25}Mn$ $1s^2, 2s^2, 2p^6, 3s^2, 3p^6, 4s^2, 3d^5$

$_{28}Ni$ $1s^2, 2s^2, 2p^6, 3s^2, 3p^6, 4s^2, 3d^8$

However, note that there are two exceptions, copper and chromium, to the pattern that two electrons always go into the 4s sublevel first.

$_{29}Cu$ $1s^2, 2s^2, 2p^6, 3s^2, 3p^6, 4s^1, 3d^{10}$

$_{24}Cr$ $1s^2, 2s^2, 2p^6, 3s^2, 3p^6, 4s^1, 3d^5$

This happens because of the stability of full and, to a lesser extent, half-filled d sublevels.

Exam tip

Remember that chromium atoms and copper atoms have only one electron in their 4s sublevel, or $4s^1$, $3d^x$. The other transition metals have two — that is, $4s^2$, $3d^x$.

Electronic configurations of ions

REVISED

Example

Write the electronic configuration of the following ions:

Fe^{3+}; Mn^{2+}; Cr^{3+}; Cu^{2+}

Answer

Step 1: write the electronic configuration of the atoms:

Fe $1s^2, 2s^2, 2p^6, 3s^2, 3p^6, 4s^2, 3d^6$

Mn $1s^2, 2s^2, 2p^6, 3s^2, 3p^6, 4s^2, 3d^5$

Cr $1s^2, 2s^2, 2p^6, 3s^2, 3p^6, 4s^1, 3d^5$

Cu $1s^2, 2s^2, 2p^6, 3s^2, 3p^6, 4s^1, 3d^{10}$

Step 2: Remove the electrons required to form the ion charge (4s out first):

Fe^{3+} $1s^2, 2s^2, 2p^6, 3s^2, 3p^6, 3d^5$

Mn^{2+} $1s^2, 2s^2, 2p^6, 3s^2, 3p^6, 3d^5$

Cr^{3+} $1s^2, 2s^2, 2p^6, 3s^2, 3p^6, 3d^3$

Cu^{2+} $1s^2, 2s^2, 2p^6, 3s^2, 3p^6, 3d^9$

Typical mistake

When writing the electronic configuration of ions, remember that the electrons are removed from the 4s sublevel first (4s electrons are first in and first out.) This is commonly forgotten in exams.

General properties of transition metals

A maximum of 10 electrons can be added to a d sublevel but, as can be seen from many of the electronic configurations described previously, many atoms and ions have a partially filled d sublevel — that is, fewer than 10 electrons.

Many of the properties of transition metal atoms and ions are explained by their partially filled d sublevels:

- formation of complexes
- formation of coloured ions
- variable oxidation states
- catalytic behaviour

Exam tip

Many of the special properties of transition metals arise from an incomplete d sublevel in atoms or ions.

Complexes

REVISED

Transition metal ions can combine with molecules or ions to form **complexes**. A complex involves one or more ligands that have formed **dative** or **coordinate** bonds to a central metal cation.

A **ligand** is a molecule or ion with the capacity to donate an electron pair to form a coordinate (dative) bond.

Some ligands form just one coordinate bond per ligand — these are called **unidentate** or **monodentate** ligands. Water, ammonia, chloride ions, cyanide ions are all monodentate ligands.

A **ligand** is a molecule or ion with the capacity to donate an electron pair to a metal ion and form a dative bond.

Other ligands can form more than one bond per ligand:

- Those that form two coordinate bonds per ligand are called **bidentate** ligands — examples (Figure 21.1) include ethane-1,2-diamine ('en'), $H_2N(CH_2)_2NH_2$ and the ethanedioate ion, $C_2O_4^{2-}$.
- Those that form more than two coordinate bonds per ligand are called **multidentate** ligands. Figure 21.2 shows a derivative of EDTA, which can form six coordinate bonds. This ligand is normally used at high pH so that all four carboxylic acid groups are deprotonated (as shown) and the ligand is $[EDTA]^{4-}$.

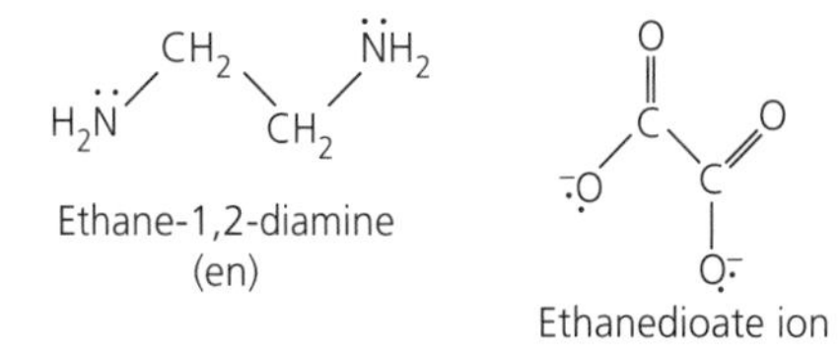

Figure 21.1 Bidentate ligands

Figure 21.2 A multidentate ligand

Typical mistake

The coordination number is *not* the same as the number of ligands. This is only true if the ligands are unidentate. The coordination number is the number of coordinate bonds formed by the transition metal ion.

Coordination number

The number of coordinate bonds formed between a metal ion and its ligands is called the ion's **coordination number**.

In the complex shown in Figure 21.3, in which four chloride ions, Cl^-, act as ligands to a central copper atom, four coordinate bonds are formed, so the coordination number is 4.

Figure 21.3 Coordination number 4

In the complex shown in Figure 21.4, there is only one ligand, $EDTA^{4-}$, but it has formed six coordinate bonds to the central metal ion, M^+. So, the coordination number is 6.

Figure 21.4 Coordination number 6

Ligands that are able to form at least two dative coordinate bonds to a transition metal are called **chelating agents**. EDTA is a good example of a chelating agent because it is able to form six coordinate bonds and 'wrap' itself around the central transition metal ion.

Small ligands like water and ammonia normally form octahedral complexes in which the coordination number of the complex is 6. In the complex in Figure 21.5, six cyanide ions, CN^-, act as ligands and form an octahedral complex with iron(III) ions.

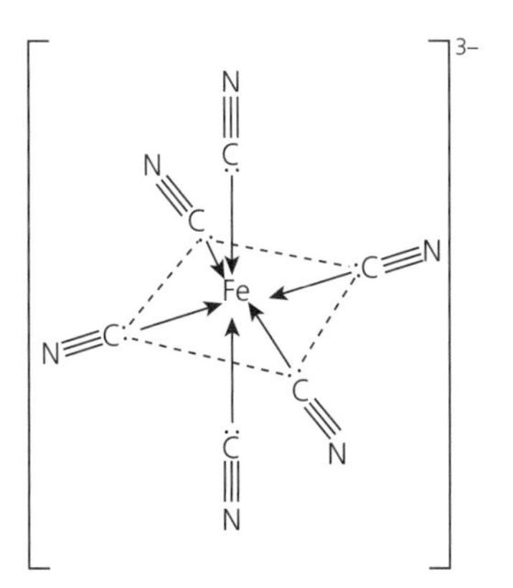

Figure 21.5 Coordination number 6

If a ligand is large then there may be only sufficient room for a maximum number of ligands. For example, the complex drawn in Figure 21.6 has a coordination number of 4 and is tetrahedral.

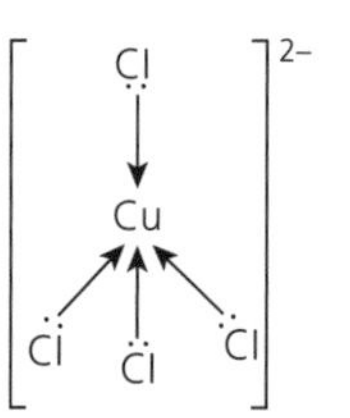

Figure 21.6 Coordination number 4

Other complexes can form with a coordination number of 4, but with the ligands arranged in a **square planar** shape, as opposed to tetrahedral. Figure 21.7 shows a complex called cis-platin (an anti-cancer compound). The complex binds with some of the nitrogen atoms in the bases within DNA and prevents the further replication that results in cancerous growths. However, the compound is highly toxic and will also affect other parts of the body if concentrations and targets are not controlled carefully.

> **Exam tip**
>
> When ligands are relatively large, fewer can 'fit' around the central metal ion.

H_3N Cl
Pt
H_3N Cl

Figure 21.7 Cis-platin

Figure 21.8 shows that the complex of a silver ion and ammonia molecules is linear, with a coordination number of 2. It is used in Tollens' reagent and forms a silver mirror when silver ions are reduced to silver atoms if warmed with an aldehyde (but not with a ketone).

Figure 21.8 The active ingredient in Tollens' reagent

Colorimetry can be used to determine the ratio of the ligands to the transition metal ion. Different ratios of the ligand solution and the transition metal ion solution are placed in a colorimeter, and the solution that has the maximum absorbance will be of the correct stoichiometric ratio of transition metal ion to ligand. Any excess ligand solution or transition metal ion solution will result in a dilution of the solution, and the absorbance will decrease.

Figure 21.9 shows haem, the compound responsible for forming the red pigment in blood. It consists of a porphyrin ring structure forming four coordinate bonds with a central iron atom. The coordination number is 4, so there are two lone pairs available for carrying oxygen.

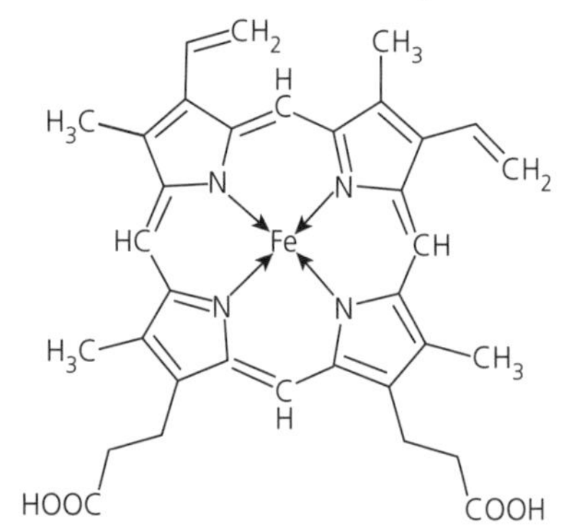

Figure 21.9 Haem — the oxygen carrier

This complex enables oxygen to be transported around the body by having available sites on the iron atom for oxygen molecules to act as ligands and be 'carried off' in the bloodstream. However, if carbon monoxide is inhaled, this molecule also acts as a ligand and binds strongly to the iron atom, forming carboxyhaemoglobin. In doing so, the available sites for oxygen transportation are reduced and this then results in less oxygen reaching important organs in the body. For this reason, carbon monoxide is toxic.

> **Exam tip**
>
> Try to learn the different shapes in terms of coordination numbers: octahedral = 6; tetrahedral and square planar = 4; linear = 2 etc.

Now test yourself

TESTED

1 What is the coordination number of the metal ion in each of these complexes?
 (a) $[CuCl_4]^{2-}$
 (b) $[Fe(H_2O)_6]^{3+}$
 (c) $[Pt(NH_3)_2Cl_2]$
 (d) $[Ni(EDTA)]^{2-}$
 (e) $[Cr(en)_3]^{3+}$
2 A complex has the formula $[Cr(C_2O_4)_3]^{3-}$.
 (a) What is the name of the ligand in the complex?
 (b) What is the shape of the complex?
 (c) What is the oxidation state of the chromium?
 (d) The same ligand reacts with nickel(II) ions. Suggest the formula of the complex formed.
3 The ligand below is called indolo[2,3-*b*]carbazole, abbreviated to 'ind'.

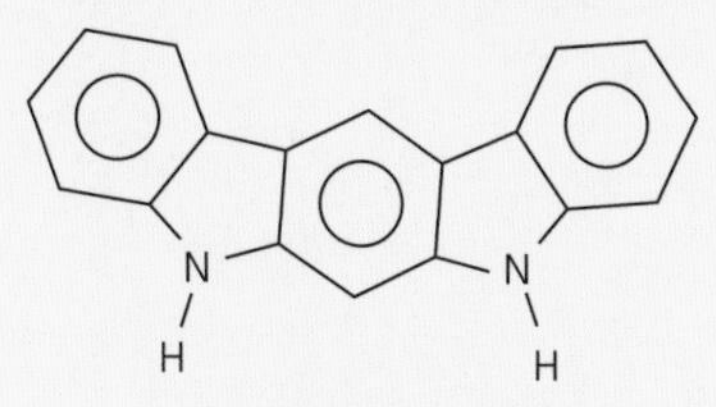

indolo[2,3-b]carbazole

 (a) How many bonds would one 'ind' molecule be expected to form with a transition metal ion?
 (b) Using the abbreviation 'ind', write the formula of the octahedral complex formed when 'ind' bonds with iron(II) ions.

Answers on p. 221

Formation of coloured ions

REVISED

Many transition metal compounds are coloured, especially in aqueous solution, because they contain coloured complexes:

- $[Co(H_2O)_6]^{2+}$ is pink.
- $[Cr(H_2O)_6]^{3+}$ is green.
- $[Fe(H_2O)_6]^{2+}$ is pale green.
- $[Fe(H_2O)_6]^{3+}$ is yellow/brown.
- $[Cu(H_2O)_6]^{2+}$ is blue.
- $[CuCl_4]^{2-}$ is yellow.
- $[CoCl_4]^{2-}$ is blue.
- $[Ag(NH_3)_2]^{+}$ is colourless.

There are several factors to consider.

- Colours might change if any or all of oxidation state, coordination number or ligand change.
- Colour arises when visible light is absorbed by a complex and electrons are promoted from the ground state to a higher energy state.
- There are five d orbitals and all are of the same energy when the transition metal is not bonded to anything else. However, when six ligands approach along the *x*, *y* and *z* axes, the lone pairs of the ligands repel the d orbitals, especially those that are pointing directly along the axes.

As there are two d orbitals along axes, and three between the axes, an energy difference now arises between the two groups of orbitals (Figure 21.10).

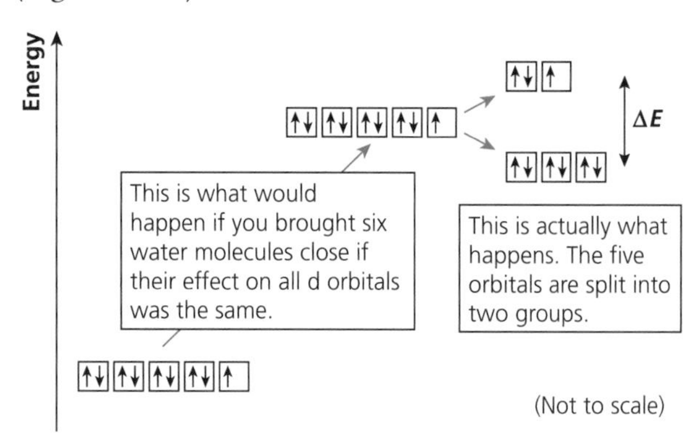

Figure 21.10

- The energy difference between the ground state and higher state, ΔE, involves a photon of frequency, ν, such that:

  $\Delta E = h\nu$

  where h is the Planck constant.
- The light that is transmitted consists of the frequencies that are not absorbed by the complex.
- For metal ions that have either an empty or completely full d sublevel, no electronic transitions are possible and therefore no absorption of visible light will occur. The resulting complex will be colourless.

The concentration of transition metal ions in solution can be determined by their absorbance of visible light. The higher the concentration of the ions, the greater the absorbance. Visible spectrometry can be used to make such a determination.

Variable oxidation states

REVISED

Many transition elements show variable oxidation states in their compounds. For example, iron can be +2 or +3, copper can be +1 or +2 and manganese can be +2, +4, +6 or +7 in their common oxidation states.

Vanadium can exist in four different oxidation states: + 5, + 4, + 3 and + 2 (Table 21.1).

Table 21.1

Oxidation state	Ion	Colour
+5	VO_3^- or VO_2^+	Yellow
+4	VO^{2+}	Blue
+3	V^{3+}	Green
+2	V^{2+}	Violet

Calculations involving redox processes

Titration calculations involving manganate(VII) ions or dichromate(VI) ions are common in examinations. Both oxidising agents can oxidise many other ions (for example chloride, Cl^-, iron(II), Fe^{2+} and iodide, I^-). It is therefore a convenient way of determining the concentrations or amounts of these ions in a sample.

Example

In an experiment to determine the percentage by mass of iron in some iron tablets, the following process was carried out.

Some iron tablets of total mass 3.67 g were ground up and dissolved in excess, warm, dilute sulfuric(VI) acid and the solution was then made up to 250 cm^3 in a volumetric flask. 25.0 cm^3 of this solution was removed by pipette and titrated against 0.0500 mol dm^{-3} potassium manganate(VII) solution. It was found that 18.50 cm^3 of the solution was required. Calculate the percentage by mass of iron in the tablets. (A_r of iron = 55.8)

Answer

$MnO_4^- + 8H^+ + 5Fe^{2+} \rightarrow 5Fe^{3+} + Mn^{2+} + 4H_2O$

amount of MnO_4^- used = $\frac{18.50}{1000} \times 0.0500 = 9.25 \times 10^{-4}$ mol

amount of Fe^{2+} reacting (from the equation) = 9.25×10^{-4} mol × 5

= 4.625×10^{-3} mol

total amount of Fe^{2+} in 250 cm^3 = $\frac{4.625 \times 10^{-3} \times 250}{25}$ = 0.04625 mol

mass of iron = 0.09625 × 55.8 = 2.58 g

percentage by mass of iron = $\frac{2.58}{3.67} \times 100$

= 70.3%

Now test yourself

TESTED

4 In an experiment to determine the quantity of iodide in a sample of dried seaweed, 50.0 g of the seaweed was treated in water and the released iodide ions were made up to 250 cm^3 in a volumetric flask. 25.0 cm^3 of the solution was removed, an indicator added and the solution was found to react with 21.25 cm^3 of 0.0100 mol dm^{-3} potassium dichromate(VI) solution.
Calculate the percentage of iodine in the seaweed. (A_r of iodine = 126.9)

Answers on pp. 221–222

Catalytic behaviour of transition metals

REVISED

There are many **catalysts** but only two types of **catalysis** involving transition metals — **homogeneous** and **heterogeneous**.

A **catalyst** is a substance that increases the rate of a chemical reaction by providing an alternative route for the reaction with a lower activation energy. At the end of the reaction, the catalyst is chemically unchanged.

Homogeneous catalysis involves a catalyst in the same physical state as the reactants.

Homogeneous catalysis

In **homogenous catalysis**, the catalyst usually forms an **intermediate species** during the reaction. Typically a transition metal ion can change its oxidation state from a higher to a lower one, and then back up to a higher state (or vice versa). The ability of transition metals to change their oxidation state in this way makes them very efficient catalysts for redox reactions.

The reaction between peroxodisulfate(VII) ions and iodide ions:

$S_2O_8^{2-}(aq) + 2I^-(aq) \rightarrow 2SO_4^{2-}(aq) + I_2(aq)$

It is normally slow because two negative ions need to interact. If iron(II) ions are added, the following processes takes place more quickly because oppositely charged ions interact in each stage.

Electrode potentials are useful in predicting what may happen in each stage:

$Fe^{3+}(aq) + e^- \rightleftharpoons Fe^{2+}(aq)$ $E^\ominus = +0.77\,V$

$I_2(aq) + 2e^- \rightleftharpoons I_2(aq)$ $E^\ominus = +0.54\,V$

$S_2O_8^{2-}(aq) + 2e^- \rightleftharpoons 2SO_4^{2-}(aq)$ $E^\ominus = 2.01\,V$

Step 1: $S_2O_8^{2-}$ ions react with Fe^{2+} ions:

$S_2O_8^{2-}(aq) + 2e^- \rightarrow 2SO_4^{2-}(aq)$ $E^\ominus = +2.01\,V$

$Fe^{2+}(aq) \rightarrow Fe^{3+}(aq) + e^-$ $E^\ominus = -0.77\,V$

Adding gives: $2Fe^{2+}(aq) + S_2O_8^{2-}(aq) \rightarrow 2SO_4^{2-}(aq) + 2Fe^{3+}(aq)$

where $E^\ominus = +2.01 + (-0.77) = +1.24\,V$

Step 2: Fe^{3+} ions react with I^- ions:

$Fe^{3+}(aq) + e^- \rightarrow Fe^{2+}(aq)$ $E^\ominus = +0.77\,V$

$2I^-(aq) \rightarrow I_2(aq) + 2e^-$ $E^\ominus = -0.54\,V$

Adding gives: $2Fe^{3+}(aq) + 2I^-(aq) \rightarrow I_2(aq) + 2Fe^{2+}(aq)$

where $E^\ominus = +0.77 + (-0.54) = +0.23\,V$

Notice how iron(II) is consumed in the first step, and then reformed in the second step.

Another reaction in which the catalyst is in the same phase as the reactants is that between ethanedioate ions and manganate(VII) ions:

$$2MnO_4^-(aq) + 16H^+(aq) + 5C_2O_4^{2-}(aq) \rightarrow 2Mn^{2+}(aq) + 8H_2O(l) + 5CO_2(aq)$$

This reaction is also very slow under normal conditions because two negative ions need to react. The reaction is catalysed by manganese(II) ions. Because these are formed in the reaction itself, the process is known as **autocatalysis**.

Heterogeneous catalysis

In **heterogenous catalysis**, the catalyst is in a phase that is different from that of the reactants.

Heterogeneous catalysis involves a catalyst in a different physical state from the reactants.

Examples of heterogeneous catalysis include:

- using iron in the Haber process:

  $$N_2(g) + 3H_2(g) \rightleftharpoons 2NH_3(g)$$

- using chromium(III) oxide in the manufacture of methanol:

  $$CO(g) + 2H_2(g) \rightarrow CH_3OH(l)$$

- using rhodium, palladium and platinum in catalytic convertors in car exhausts:

  $$2CO(g) + 2NO(g) \rightarrow CO_2(g) + N_2(g)$$

- using vanadium(V) oxide in the Contact process:

  $$2SO_2(g) + O_2(g) \rightleftharpoons 2SO_3(g)$$

In this type of catalysis, the reacting molecules are **adsorbed** onto the surface of the catalyst. The reaction then takes place on the surface. Clearly, maximising the surface area by providing a support medium will increase the rate of reaction further and reduce potential costs. For example, rhodium is used on a ceramic support in catalytic convertors in cars.

Sometimes, for example in the Haber process and in catalytic convertors, the catalyst may become **poisoned**. Foreign atoms bond to the active sites on the surface of the catalyst, reducing the effective surface area available for reaction. This means that there will be costs incurred due to the inefficient operation of catalysts.

In the Contact process, vanadium(v) oxide provides a surface on which the reaction takes place, but it also reacts by changing the oxidation state of the vanadium, forming an intermediate, V_2O_4:

$$V_2O_5(s) + SO_2(g) \rightarrow V_2O_4(s) + SO_3(g)$$

$$V_2O_4(s) + \frac{1}{2}O_2(g) \rightarrow V_2O_5(s)$$

Overall:

$$SO_2(g) + \frac{1}{2}O_2(g) \rightarrow SO_3(g)$$

A **Lewis acid** is a lone-pair acceptor. A **Lewis base** is a lone-pair donor.

Reactions of hydrated transition metal ions

When ligands bond to transition metal ions, the ligands act as **Lewis bases** — they donate lone pairs to form coordinate bonds. Transition metal ions accept lone pairs and are therefore **Lewis acids**.

Figure 21.11 shows a typical hydrated complex, which often involves six water molecules forming coordinate bonds to a central transition metal ion, M^{n+}, to form an octahedral complex.

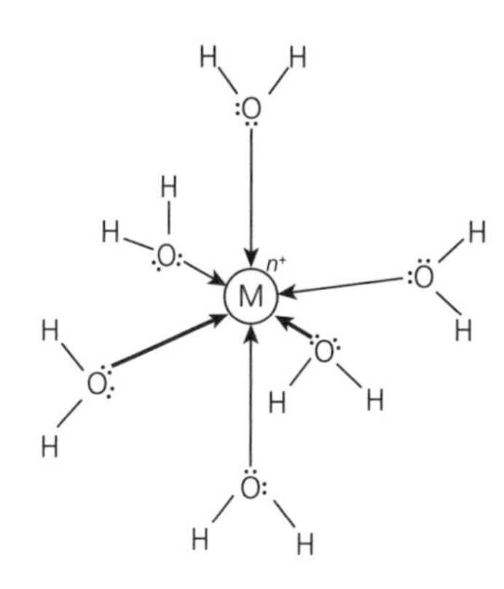

Figure 21.11 Typical octahedral aqua-complex

Aqueous ions in solution

REVISED

Many transition metals form aqueous ions with charges of +2 or +3:

+2 ions: $[Fe(H_2O)_6]^{2+}$ $[Co(H_2O)_6]^{2+}$ $[Cu(H_2O)_6]^{2+}$

+3 ions: $[Fe(H_2O)_6]^{3+}$ $[Al(H_2O)_6]^{3+}$ $[Cr(H_2O)_6]^{3+}$

When these ions are in solution, equilibria exist in which the complex ion undergoes a hydrolysis process (effectively the loss of a proton, H^+). For example:

$$[Fe(H_2O)_6]^{2+} + H_2O \rightleftharpoons [Fe(H_2O)_5(OH)]^{+} + H_3O^{+}$$

$$[Fe(H_2O)_6]^{3+} + H_2O \rightleftharpoons [Fe(H_2O)_5(OH)]^{2+} + H_3O^{+}$$

- A +3 metal ion is smaller and more highly charged than a +2 ion, so it has a higher charge density.
- A +3 metal ion polarises water molecules to a greater extent than a +2 ion.
- This results in bond weakening of O–H bonds in the +3 aqueous complex.
- As a result, +3 aqueous ions are more acidic than +2 ions.

So, the equilibrium relating to the +3 ion has an equilibrium position that lies more to the right-hand side than the corresponding +2 equilibrium position.

As a result, a solution of +3 aqueous ions has a lower pH than a solution of +2 ions.

Transition aqueous metal ions in the +2 oxidation state

- All metals that form M^{2+} ions have +2 as the most common oxidation state (by loss of both s electrons).
- All the aqueous ions have the form of octahedral $[M(H_2O)_6]^{2+}$.
- Some common metal(II) aqueous ions colours are: Fe^{2+}, pale green; Co^{2+}, pink; Cu^{2+}, blue.

Reactions of +2 aqueous ions

With sodium hydroxide solution

All aqueous ions react with hydroxide ions and form insoluble and coloured metal(II) hydroxides as a precipitate.

Aqueous ion	Fe^{2+}	Co^{2+}	Cu^{2+}
Name of precipitate	Iron(II) hydroxide	Cobalt(II) hydroxide	Copper(II) hydroxide
Formula of precipitate	$Fe(OH)_2$	$Co(OH)_2$	$Cu(OH)_2$
Colour of precipitate	Dark green	Blue/green	Light blue

The reaction taking place is between hydroxide ions, OH^-, and the aqueous transition metal complex:

$$[M(H_2O)_6]^{2+}(aq) + 2OH^-(aq) \rightarrow [M(H_2O)_4(OH)_2](s) + 2H_2O(l)$$

or simply:

$$M^{2+}(aq) + 2OH^-(aq) \rightarrow M(OH)_2(s)$$

Ammonia solution

Ammonia is a weak base in aqueous solution and contains hydroxide ions, OH^-:

$$NH_3(aq) + H_2O(aq) \rightleftharpoons NH_4^+(aq) + OH^-(aq)$$

The hydroxide ions remove protons from the aqueous +2 complex in the first step:

$$[M(H_2O)_6]^{2+}(aq) + 2OH^-(aq) \rightarrow [M(H_2O)_4(OH)_2](s) + 2H_2O(l)$$

In some cases, the hydroxide precipitate dissolves in excess ammonia solution. In this process, ammonia, NH_3, acts as a ligand and removes the water molecules and hydroxide ions from the original complex in a process known as **ligand substitution**:

$$[M(H_2O)_4(OH)_2](s) + 6NH_3(aq) \rightleftharpoons [M(NH_3)_6]^{2+}(aq) + 4H_2O(l) + 2OH^-(aq)$$

- Iron(II) aqueous ions form iron(II) hydroxide in the first stage, but no further ligand substitution takes place.
- Cobalt(II) aqueous ions form a dark blue precipitate of cobalt(II) hydroxide; this dissolves in excess ammonia to form the straw-coloured cobalt(II)–ammonia complex $[Co(NH_3)_6]^{2+}$.

- Copper(II) aqueous ions form a light blue precipitate of copper(II) hydroxide; this dissolves in excess ammonia to form the dark blue complex $[Cu(NH_3)_4(H_2O)_2]^{2+}$.

Sodium carbonate solution

The carbonate ions, CO_3^{2-}, react with aqueous 2^+ ions to form insoluble carbonate precipitates:

$$M^{2+}(aq) + CO_3^{2-}(aq) \rightarrow MCO_3(s)$$

Name of precipitate	Iron(II) carbonate	Cobalt(II) carbonate	Copper(II) carbonate
Formula of precipitate	$FeCO_3$	$CoCO_3$	$CuCO_3$
Colour of precipitate	Green	Pink	Green-blue

Transition metal ions in the +3 oxidation state

In aqueous solution, they all exist as the $[M(H_2O)_6]^{3+}$ ion, although **hydrolysis** always occurs to form the $[M(H_2O)_5(OH)]^{2+}$ ion along with protons — so solutions containing these ions are weak acids. Aqueous solutions of Cr^{3+}, Fe^{3+} and Al^{3+} (for comparison) tend to be acidic and have pK_a values similar to that of ethanoic acid.

Exam tip

Many of the reactions of +3 aqueous ion solutions are those that an acid would undergo, particularly when reacting with carbonates to form carbon dioxide gas.

Reactions of aqueous +3 ions

Sodium hydroxide solution

The equilibrium written below shifts to the right-hand side on adding hydroxide ions, OH^-, by reacting with hydrogen ions:

$$[Fe(H_2O)_6]^{3+}(aq) + H_2O(l) \rightleftharpoons [Fe(H_2O)_5(OH)]^{2+}(aq) + H_3O^+(aq)$$

or

$$[Fe(H_2O)_6]^{3+}(aq) + OH^-(aq) \rightarrow [Fe(H_2O)_5(OH)]^{2+}(aq) + H_2O(l)$$

This process repeats itself until the metal(III) hydroxide precipitate (with zero charge) forms.

- Iron(III) aqueous ions form a brown precipitate of hydrated iron(III) hydroxide:

$$[Fe(H_2O)_6]^{3+}(aq) + 3OH^-(aq) \rightarrow [Fe(OH)_3(H_2O)_3](s) + 3H_2O(l)$$

- Chromium(III) aqueous ions form a green precipitate of chromium(III) hydroxide, but this dissolves in excess hydroxide ions to form a green solution containing $[Cr(OH)_6]^{3-}$ ions:

$$[Cr(H_2O)_6]^{3+}(aq) + 3OH^-(aq) \rightarrow [Cr(OH)_3(H_2O)_3](s) + 3H_2O(l)$$

then:

$$[Cr(OH)_3(H_2O)_3](s) + 3OH^-(aq) \rightarrow [Cr(OH)_6]^{3-}(aq) + 3H_2O(l)$$

- Aluminium aqueous ions have a similar reaction to that of chromium(III) ions, although the hydroxide precipitate is white and a colourless solution forms on adding excess sodium hydroxide solution:

$$[Al(H_2O)_6]^{3+}(aq) + 3OH^-(aq) \rightarrow [Al(OH)_3(H_2O)_3](s) + 3H_2O(l)$$

then:

$$[Al(OH)_3(H_2O)_3](s) + OH^-(aq) \rightarrow [Al(OH)_4(H_2O)_2]^-(aq) + H_2O(l)$$

The hydroxides of aluminium and chromium(III) are **amphoteric** because they react with both acids and bases to form salts.

Ammonia solution

Ammonia is a weak base in aqueous solution and contains hydroxide ions, OH^-:

$$NH_3(aq) + H_2O(aq) \rightleftharpoons NH_4^+(aq) + OH^-(aq)$$

The hydroxide ions form the metal(III) hydroxides as precipitates — the same as those produced using sodium hydroxide solution (Table 21.2).

Table 21.2

Name of precipitate	Iron(III) hydroxide	Chromium(III) hydroxide	Aluminium hydroxide
Formula of precipitate	$Fe(OH)_2$	$Cr(OH)_3$	$Al(OH)_3$
Colour of precipitate	Dark green	Green	White

When excess ammonia solution is added:

- iron(III) hydroxide precipitate is unaffected
- chromium(III) hydroxide dissolves slowly to form the violet complex ion $[Cr(NH_3)_6]^{3+}$
- aluminium hydroxide precipitate is unaffected

Sodium carbonate solution

An aqueous solution of a carbonate is alkaline:

$$CO_3^{2-}(aq) + H_2O(l) \rightleftharpoons HCO_3^-(aq) + OH^-(aq)$$

So a reaction is expected in which metal(III) hydroxide precipitates form.

However, carbonate ions will also react with the acidic aqueous metal(III) ions to form carbon dioxide gas:

$$CO_3^{2-}(aq) + H_3O^+(aq) \rightarrow CO_2(g) + H_2O(l)$$

In the case of chromium(III) ions in solution, this reaction takes place:

$$2[Cr(H_2O)_6]^{3+}(aq) + 3CO_3^{2-}(aq) \rightarrow 2[Cr(H_2O)_3(OH)_3](s) + 3CO_2(g) + 3H_2O(l)$$

$[Fe(H_2O)_6]^{3+}(aq)$ and $[Al(H_2O)_6]^{3+}(aq)$ react in the same way as $[Cr(H_2O)_6]^{3+}(aq)$.

Exam tip

Remember that aqueous solutions containing +3 metal ions are acidic. They will react with sodium carbonate to form carbon dioxide gas, and so fizzing will be seen.

Typical mistake

Do not make the error of stating that a metal(III) carbonate forms when a carbonate is added to an aqueous metal(III) ion solution. The aqueous ion is too acidic and decomposes the carbonate ion to form carbon dioxide gas — so a metal(III) hydroxide forms instead.

Now test yourself

TESTED

5 State what is observed and write equations for all reactions in each of the following:
  (a) ammonia solution is added dropwise until in excess to aqueous copper(II) ions
  (b) sodium hydroxide solution is added dropwise until in excess to chromium(III) ions
  (c) sodium carbonate solution is added to aqueous aluminium(III) sulfate solution

6 Explain how sodium carbonate solution can be used to distinguish between a solution containing iron(II) ions and one containing iron(III) ions.

Answers on p. 222

Ligand substitution reactions

Ligand substitution is a process that can happen to a complex ion, depending on conditions.

Ligand substitution reactions happen when one ligand in a complex is replaced by another.

Ligands with molecules of similar size — for example ammonia and water — can be interchanged without any change of coordination number:

$$[Co(H_2O)_6]^{2+}(aq) + 6NH_3(aq) \rightleftharpoons [Co(NH_3)_6]^{2+}(aq) + 6H_2O(l)$$

$$[Cr(H_2O)_6]^{3+}(aq) + 6NH_3(aq) \rightleftharpoons [Cr(NH_3)_6]^{3+}(aq) + 6H_2O(l)$$

You can see that the coordination number of all the complexes remains at 6.

As mentioned in an earlier section, aqueous copper(II) ions react with ammonia ligands, but the substitution is not complete — only four ammonia ligands bond to the copper(II) ion instead of six:

$$[Cu(H_2O)_6]^{2+}(aq) + 4NH_3(aq) \rightleftharpoons [Cu(NH_3)_4(H_2O)_2]^{2+}(aq) + 4H_2O(l)$$

So ligand substitution has happened but the coordination number is still 6.

Ligands like chloride ions, Cl^-, are often relatively large and are also charged. This means that geometric and electrostatic repulsion between the ligands takes place when they are bonding to a transition metal ion and, as a result, fewer charged ligands can bond to the metal ion. This can result in a change in coordination number — from 6 to 4 in the examples below:

$$[Cu(H_2O)_6]^{2+}(aq) + 4Cl^-(aq) \rightleftharpoons [CuCl_4]^{2-}(aq) + 6H_2O(l)$$

$$[Co(H_2O)_6]^{2+}(aq) + 4Cl^-(aq) \rightleftharpoons [CoCl_4]^{2-}(aq) + 6H_2O(l)$$

When multidentate ligands are used — for example ethane-1,2-diamine, ethanedioate ions (both bidentate ligands) or EDTA (hexadentate) — then stable complexes are formed:

$$[Cu(H_2O)_6]^{2+}(aq) + EDTA^{4-}(aq) \rightleftharpoons [Cu(EDTA)]^{2-}(aq) + 6H_2O(l)$$

The **chelating effect** of using a ligand like EDTA results in a highly stable complex. One of the reasons is that there is a large increase in entropy when the reaction takes place, due to many water molecules being displaced:

$$[Cr(H_2O)_6]^{2+}(aq) + EDTA^{4-}(aq) \rightleftharpoons [Cr(EDTA)]^{2-}(aq) + 6H_2O(l)$$

There are two species on the left-hand side of the equilibrium and seven species on the right-hand side. There will therefore be an increase in entropy of the system. This will decrease the value for the Gibbs free energy, and hence make the complex more thermodynamically stable.

Now test yourself

TESTED

7 This ligand is known by the abbreviation 'TPEDA':

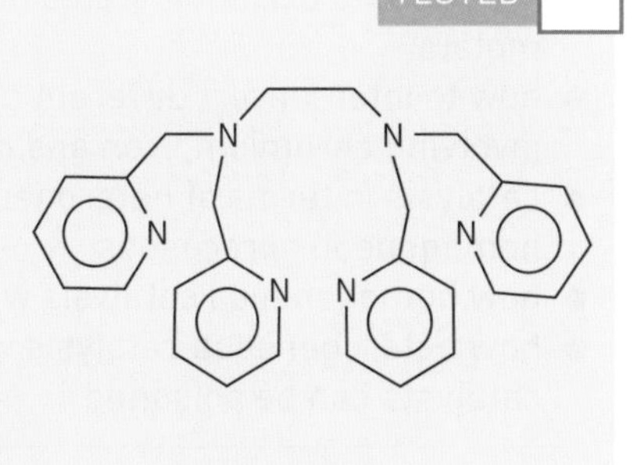

(a) Suggest how many coordinate bonds TPEDA will form with a transition metal ion.

(b) Write an equation to show how TPEDA would react with aqueous iron(II) ions.

(c) Suggest why the iron(II)–TPEDA complex is thermodynamically stable.

Answers on p. 222

Exam practice

1 (a) In terms of bonding, explain the term 'complex'. [2]

(b) Identify *one* of the species in this list that does *not* act as a ligand. Explain your answer. [2]

CN^- O^{2-} CH_4 NH_3

(c) The element gold is in the d-block of the periodic table. Consider the following gold complex (known as 'audien'), which contains a chloride ion, Cl^-:

$[$NH; NH_2—Au—NH_2; Cl$]^{2+}$

(i) Deduce the oxidation state of gold in this complex ion. [1]

(ii) Give the coordination number of gold in this complex ion. [1]

(iii) Give the names of two possible shapes for audien. [2]

2 This question is about cobalt chemistry.

(a) Aqueous cobalt(II) ions, $[Co(H_2O)_6]^{2+}(aq)$, are pink.

(i) With reference to electrons, explain why aqueous cobalt(II) ions are pink. [3]

(ii) By reference to aqueous cobalt(II) ions, state the meaning of each of the terms in the equation $\Delta E = h\nu$. [3]

(iii) Write an equation for the reaction, in aqueous solution, between $[Co(H_2O)_6]^{2+}$ and an excess of chloride ions. [2]

(iv) State the shape of the complex produced in part (a)(iii) and explain why its shape differs from that of the $[Co(H_2O)_6]^{2+}$ ion. [3]

(v) Draw the structure of the ethanedioate ion ($C_2O_4^{2-}$). Explain how this ion can act as a ligand. [2]

(b) When a dilute solution containing ethanedioate ions is added to a solution containing cobalt(II) ions, a substitution reaction occurs. In this reaction, four water molecules are replaced and a new complex is formed.

(i) Write an ionic equation for the reaction. Give the coordination number of the complex formed and name its shape. [4]

(ii) In the complex formed, the two water molecules are opposite each other. Draw a diagram to show how the ethanedioate ions are bonded to the cobalt(II) ion and give a value for the O–Co–O bond angles. Don't show the water molecules in your diagram. [2]

Answers and quick quiz 21 online

ONLINE

Summary

You should now have an understanding of:

- the general properties of transition metals
- complexes
- ligands and the different types that exist
- the coordination number of a complex
- the various shapes of complexes
- how colour arises in a complex
- the variable oxidation states of transition metals
- how to interconvert different oxidation states involving chromium, iron and cobalt
- catalysis in terms of heterogeneous and homogeneous processes
- how homogeneous catalysis works
- how heterogeneous catalysis works and how catalysts can be poisoned
- various applications of transition metal catalysts
- Lewis acids and bases
- the existence of transition metal +2 and +3 aqueous ions
- the relative acidity of +2 and +3 ions in terms of the charge density of the positive ion
- the reactions of various +2 and +3 ions with ammonia, hydroxide ions and carbonate ions
- ligand substitution reactions and the factors that affect the coordination numbers in different complexes
- how the thermodynamic stability of complexes can be related to entropy effects during ligand substitution

22 Isomerism in organic chemistry

Structural isomers

Structural isomers are defined as compounds that have the same molecular formula but different structures. For example, the structural isomers of C_5H_{12} with different carbon chains are shown in Figure 22.1. They are called pentane, 2-methylbutane and 2,2-dimethylpropane respectively.

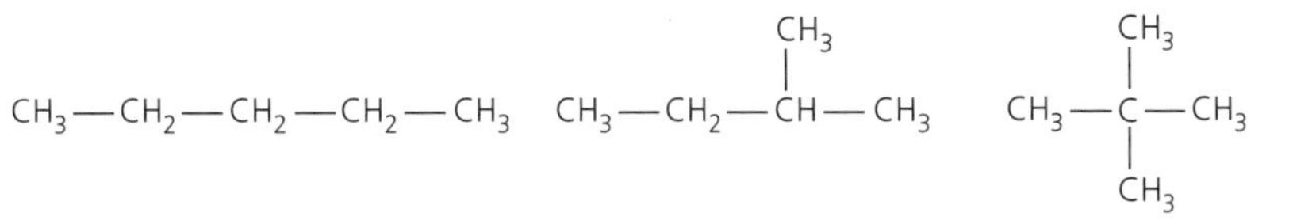

$CH_3—CH_2—CH_2—CH_2—CH_3$ $CH_3—CH_2—CH(CH_3)—CH_3$ $CH_3—C(CH_3)_2—CH_3$

Pentane 2-Methylbutane 2,2-Dimethylpropane

Figure 22.1 Isomers of C_5H_{12}

Figure 22.2 shows two molecules that are also structural isomers, but these have different functional groups. The first is an alcohol and the second is an ether. Notice how both molecules consist of atoms, some of which are bonded to *different 'neighbouring' atoms*.

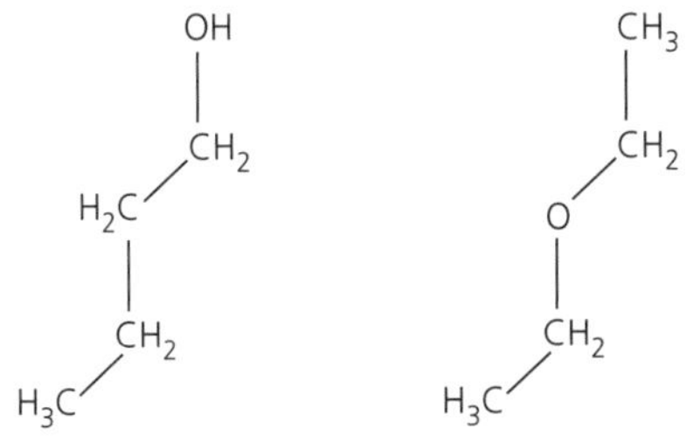

$H_3C—CH_2—H_2C—CH_2—OH$ $H_3C—CH_2—O—CH_2—CH_3$

Butan-1-ol Ethoxyethane

Figure 22.2 Functional group isomers of $C_4H_{10}O$

Structural isomers have different physical and chemical properties because their structures are different, and so their intermolecular forces differ in type and in strength.

Stereoisomerism

Stereoisomers are compounds that have the same structural formula, but their atoms differ in spatial arrangement.

- Atoms in molecules of this type are bonded to the *same 'neighbouring' atoms*, but their 3-dimensional arrangement is different.
- There are two types of stereoisomerism — ***E–Z* isomerism** and **optical isomerism**.

E–Z isomerism

REVISED

This type of isomerism was discussed in Chapter 12. It exists due to the restricted rotation about a carbon–carbon double bond. The π bonding electrons prevent free rotation, and this means that two groups, or substituents, may either be on the same side of the double bond (*Z*-isomer) or on opposite sides (*E*-isomer).

For example, Figure 22.3 shows the isomerism of but-2-ene, in which the methyl groups either side of the carbon–carbon double bond are on opposite sides (*E*) or on the same side (*Z*).

Figure 22.4 shows another example of *E–Z* isomerism using 1,2-dichloroethene.

The boiling points of the dichloro-isomers are 60.3°C and 47.5°C respectively. In the case of *Z*-1,2-dichloroethene, the $^{\delta+}C-Cl^{\delta-}$ dipoles do not cancel out, so this molecule has an overall dipole. In the case of *E*-1,2-dichloroethene, the $^{\delta+}C-Cl^{\delta-}$ dipoles do cancel and so this molecule is non-polar. The intermolecular forces (dipole–dipole forces) between *Z*-1,2-dichloroethene molecules will be stronger than the van der Waals forces between molecules of *E*-1,2-dichloroethene, and the boiling point of the *E*-isomer will be lower.

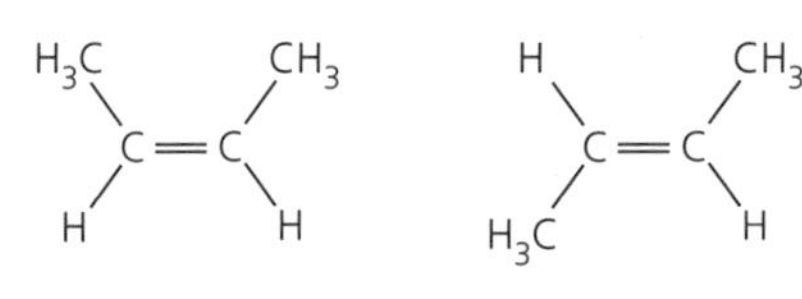

Figure 22.3 *E–Z* isomers of but-2-ene

H H C=C Cl Cl — H Cl C=C Cl H

Z-1,2-dichloroethene *E*-1,2-dichloroethene

Figure 22.4 *E–Z* isomers of 1,2-dichloroethene

Optical isomerism

REVISED

Because molecules are 3-dimensional, many are different compared with their mirror image molecules. Molecules like this that have non-superimposable mirror images are called **chiral molecules** or **enantiomers**.

One of the enantiomers will *rotate the plane of polarised light* to the right, and the other will rotate it to the left. The signs (+) and (−) are used to show that a molecule rotates plane-polarised light to the right or the left respectively.

When carbon atoms form bonds to four different groups arranged as a tetrahedron in space (or have an **asymmetric carbon atom**), enantiomers will form.

The molecules in Figure 22.5 are mirror images of each other, but they are non-superimposable. These are therefore different molecules, called optical isomers.

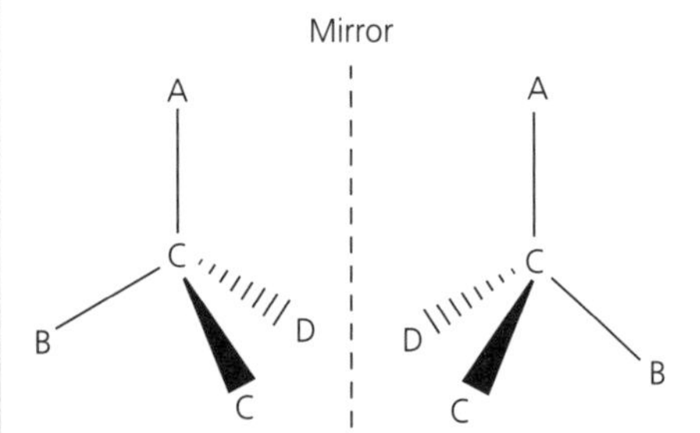

Figure 22.5 Optical isomers

Figure 22.6 shows the amino acid called alanine (2-aminopropanoic acid), with the structural formula $H_2NCH(CH_3)COOH$. Again, the mirror image molecules cannot be superimposed on each other, so alanine exists as two enantiomers.

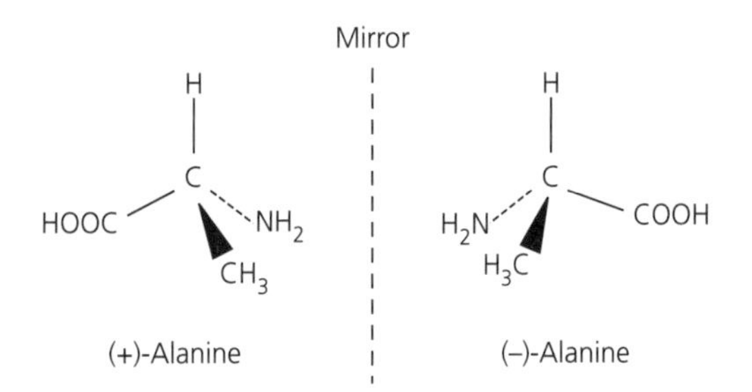

Figure 22.6 Optical isomerism in amino acids

Exam tip

When drawing the structures of enantiomers, make sure that you use 3-dimensional diagrams.

Enantiomers in reactions

REVISED

Many pharmaceutical drugs consist of molecules that are chiral, and their chirality is an essential feature of their chemical behaviour.

Often one of the enantiomeric forms has a markedly different chemical behaviour from the other form. This can be illustrated using the example of thalidomide — a drug used in the 1960s to cure nausea in pregnancy. The structures of the mirror image molecules are shown in Figure 22.7.

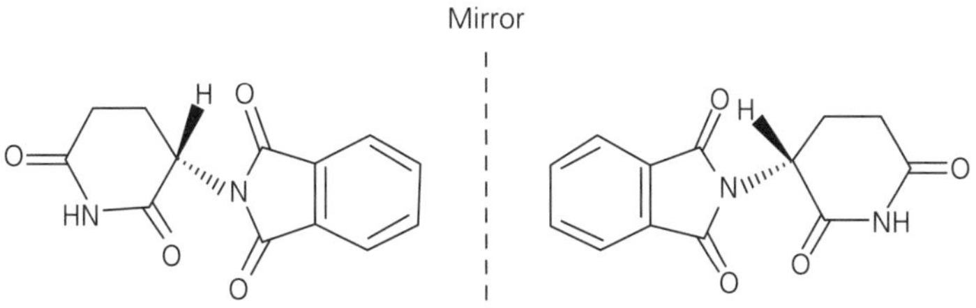

Mirror

Figure 22.7 Thalidomide

The molecule on the left reduces morning sickness — the desired effect — and the molecule on the right results in deformation of the foetus. Although the molecules look similar, subtle differences in their stereochemistry cause a huge difference to their chemical behaviour.

In this case, it is necessary to separate one enantiomer from the other before distribution for use by pregnant women. This can be expensive because it often requires the use of an extra enzymic process to remove the unwanted molecules from the reaction mixture.

Alternatively, the synthesis of a molecule can have its stereochemistry controlled so that only one of the enantiomers is formed. This technique often uses chiral reagents and can be time consuming.

In a reaction that does not have its stereochemistry controlled, it is common to produce a mixture of both the (+) and (−) optical isomers — this is called a racemic mixture or racemate. The mixture does not affect plane-polarised light because each enantiomer has an equal and opposite effect on the optical rotation of the plane-polarised light.

In many reaction mechanisms, it may be equally possible for an attacking species to react on either side of an intermediate — a species that is temporarily formed during a reaction and has a short half-life. As the reaction then takes place, it may be possible to form an equal mixture of both the (+) and (−) forms.

Now test yourself

TESTED

1 (a) What is meant by 'structural isomers'?
  (b) Draw the structural formulae and name the structural isomers of C_6H_{14}.

2 (a) Write the molecular formula of the fatty acid shown here: 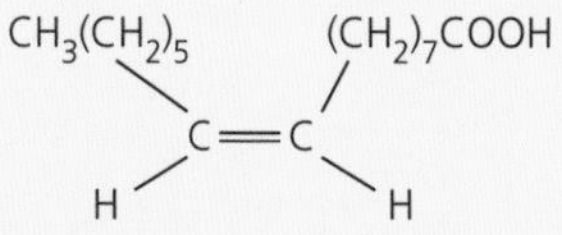
  (b) The molecule in part (a) can show *E–Z* isomerism. Explain why.
  (c) The isomer shown in the diagram is the *Z*-isomer. Draw the structure of the *E*-isomer.

$CH_3(CH_2)_5$ — C=C — $(CH_2)_7COOH$, H, H

3 (a) What is meant by the term 'stereoisomers'?
  (b) Lactic acid has the structural formula $HOOCCH(OH)CH_3$. Draw the enantiomeric forms of lactic acid.
  (c) Describe a technique by which the two enantiomeric forms of lactic acid can be distinguished.

Answers on p. 222

Exam practice

1 (a) Name molecule A. [1]

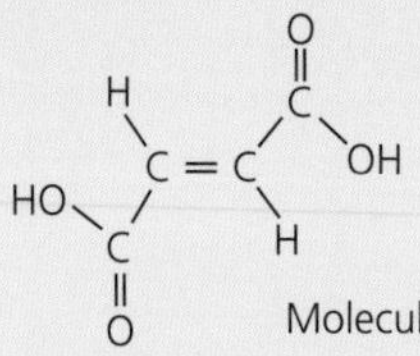

Molecule A

(b) What type of stereoisomerism can molecule A display? Explain your answer. [2]

(c) Draw the displayed formula of the other stereoisomer of A. [1]

(d) These stereoisomers have different melting points — state which has the higher melting point. Explain your answer. [2]

Answers and quick quiz 22 online

ONLINE

Summary

You should now have an understanding of:

- structural isomers
- *E–Z* isomerism
- optical isomerism
- the meaning of the terms 'enantiomer' and 'racemate'
- how to draw various isomers
- why the formation and use of specific enantiomers is important, especially in medicine

23 Compounds containing the carbonyl group

Aldehydes and ketones

Aldehydes and ketones are described as **carbonyl compounds** because they contain the carbonyl group (Figure 23.1).

Figure 23.1 The carbonyl group

Figure 23.2 shows the functional groups in an aldehyde and a ketone.

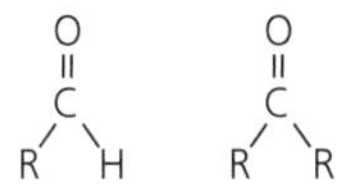

Aldehyde Ketone

Figure 23.2 Two types of carbonyl functional groups

The molecule in Figure 23.3 is called ethanal — it is an aldehyde because it contains the −CHO functional group.

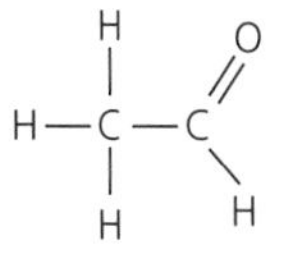

Figure 23.3 Ethanal

Other more complex, naturally occurring aldehydes and ketones are shown in Figure 23.4.

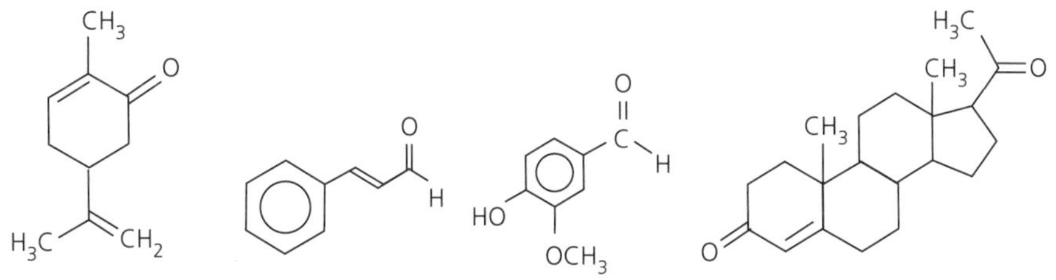

Carvone (spearmint and caraway) | Cinnamaldehyde (cinnamon bark) | Vanillin (vanilla bean) | Progesterone (female sex hormone)

Figure 23.4 Some naturally occurring carbonyl compounds

Now test yourself

TESTED

1 Which of the molecules in Figure 23.4 are aldehydes and which are ketones?

Answers on p. 222

Reactions of aldehydes and ketones

REVISED

Redox processes

At AS, one method (page 94) for synthesising aldehydes and ketones involves oxidising alcohols using acidified potassium dichromate(VI). Primary alcohols are oxidised to form aldehydes and then carboxylic acids. Secondary alcohols are oxidised to form ketones. For example, using propan-1-ol, $CH_3CH_2CH_2OH$:

$$CH_3CH_2CH_2OH + [O] \rightarrow CH_3CH_2CHO + H_2O$$

Propan-1-ol Propanal

Then, if sufficient oxidising agent [O] is present:

$CH_3CH_2CHO + [O] \rightarrow CH_3CH_2COOH$

Propanal Propanoic acid

Aldehydes are easily oxidised to form the corresponding carboxylic acid. This means that aldehydes are very good **reducing agents** and will, therefore, reduce other species, for example Ag^+ to Ag, or Cu(II) compounds to Cu(I) compounds.

Fehling's solution

Fehling's solution contains copper(II) ions. These are reduced to form copper(I) compounds such as copper(I) oxide. When an aldehyde is added to Fehling's solution and the mixture is warmed, a *red precipitate* is formed:

$2Cu^{2+}(aq) + H_2O(l) + 2e^- \rightarrow Cu_2O(s) + 2H^+(aq)$

The aldehyde is oxidised to form a carboxylic acid, for example:

$CH_3CHO(aq) + [O] \rightarrow CH_3COOH(aq)$

$CH_3CHO(aq) + H_2O(l) \rightarrow CH_3COOH(aq) + 2H^+(aq) + 2e^-$

Fehling's solution and **Tollens' reagent** can be used to distinguish between an aldehyde and a ketone. Ketones do not react with these reagents.

Tollens' reagent

Tollens' reagent contains the complex ion $[Ag(NH_3)_2]^+$. It is formed by adding ammonia solution dropwise to silver nitrate solution until the brown precipitate of silver oxide just dissolves to form a colourless solution.

When this is warmed with an aldehyde, a *silver mirror* is formed. The silver ion is reduced in the reaction:

$Ag^+(aq) + e^- \rightarrow Ag(s)$

The aldehyde is oxidised in the reaction. Ketones are not easily oxidised and so do not react.

Nucleophilic addition reactions

The carbonyl group, C=O, is unsaturated, so addition reactions can happen. The carbon atom of the group is less electronegative than the oxygen atom. The bond is therefore polarised and the carbon atom has a δ+ charge (Figure 23.5), so **nucleophiles** attack it. Hence, carbonyl compounds can take part in nucleophilic addition reactions.

A **nucleophile** seeks out centres of positive charge and is a lone-pair donor.

O δ–
C δ+

Figure 23.5 The polar carbonyl group

The nucleophile performs an addition process on the carbonyl group. The general type of nucleophilic addition is shown in Figure 23.6.

δ+ δ–
C = O
:Nu⁻
Nu—C—O⁻:
H⁺
(Tetrahedral intermediate)
H_3O^+
Nu—C—O—H
(Addition product)

Figure 23.6 Nucleophilic addition mechanism

Reduction with sodium tetrahydridoborate(III), $NaBH_4$

$NaBH_4$ in aqueous solution behaves as a reducing agent. It provides hydride ions, H^-, as nucleophiles and these react with carbonyl groups in both aldehydes and ketones to form the corresponding primary and secondary alcohols respectively.

The mechanism is a nucleophilic addition process (Figure 23.7):

- The hydride ion, H^-, donates its lone pair to the slightly positively charged carbon atom in the carbonyl group.
- The higher-energy, π-bonding electrons in the carbonyl group move onto the oxygen atom to form a negative charge.
- One of the lone pairs of electrons on the oxygen atom is used to allow protonation, normally by an acid. This forms a hydroxyl group, OH.

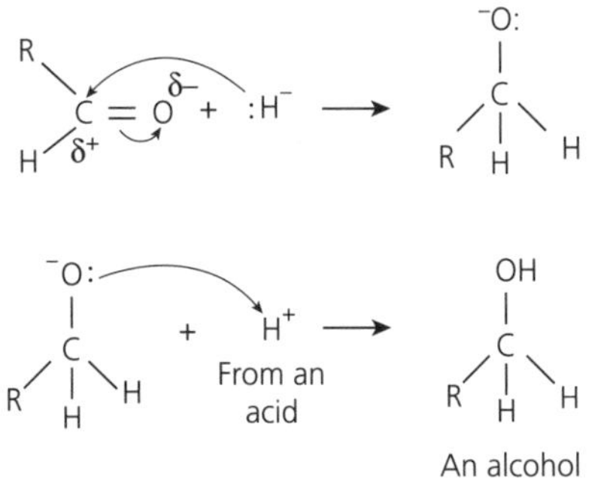

From an acid

An alcohol

Figure 23.7 **Nucleophilic addition mechanism reducing an aldehyde**

Reaction with hydrogen cyanide, HCN

It is possible to use the cyanide ion, CN^-, as a nucleophile when reacting with a carbonyl group. It reacts in the same way as the hydride ion, H^-.

The cyanide ion, CN^-, attacks the carbonyl carbon atom in this process to form a new substance called a **hydroxynitrile**.

Exam tip

Using hydrogen cyanide and cyanides (in general) is very hazardous due to the highly toxic nature of these substances.

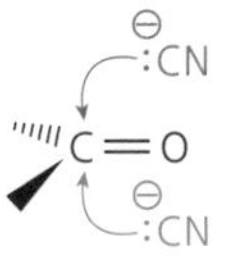

Using ethanal as an example:

$$CH_3CHO + HCN \rightarrow CH_3CH(OH)(CN)$$

HCN is a very weak acid and so provides few CN^- ions — a trace of soluble KCN provides the nucleophile.

In the first stage of this mechanism (Figure 23.8), the cyanide ion can approach the trigonal planar carbonyl group from either side (Figure 23.9).

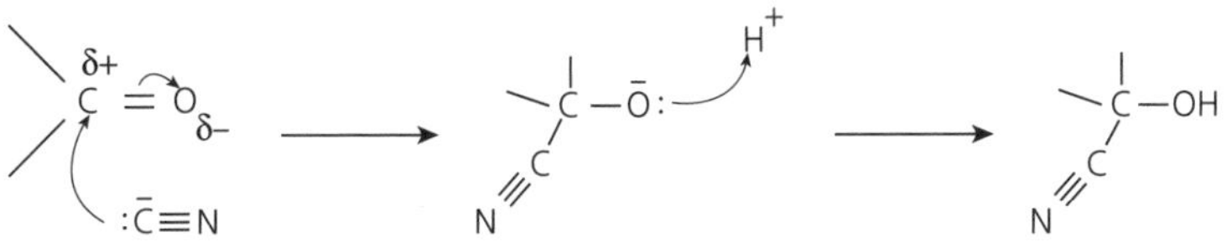

Figure 23.8

Figure 23.9

This means that the stereochemistry is not controlled and the resulting hydroxynitrile is likely to form as a 50:50 racemic mixture of enantiomers.

Now test yourself

TESTED

2 Draw the structures of the products formed when the following are reduced using $NaBH_4$.

(a)

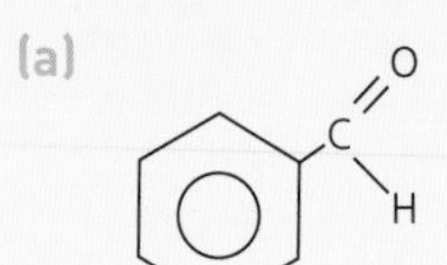

(b) $(H_3C)_2CH—CHO$

(c) $CH_3—CH_2—CH_2—CHO$

3 Draw the structure of the product formed when the following reacts with hydrogen cyanide (HCN) in KCN.

Answers on p. 222

Carboxylic acids and esters

Carboxylic acids are molecules containing the functional group shown in Figure 23.10.

—COOH or —C(=O)—O—H

Figure 23.10 Carboxylic acid functional group

The carbonyl group, C=O, withdraws electrons from the hydroxyl group, making the O–H bond weaker. This induces dissociation to form the **carboxylate anion** and a hydrated proton — for example:

$$CH_3COOH(aq) + H_2O(l) \rightarrow CH_3COO^-(aq) + H_3O^+(aq)$$

Carboxylic acids are **weak acids** because they dissociate only partially in solution.

Chemical reactions of carboxylic acids

REVISED

Acidic behaviour

Carboxylic acids show all the normal reactions of acids:

- With **reactive metals** to form a carboxylate salt and hydrogen:

  $Na(s) + CH_3COOH(aq) \rightarrow CH_3COO^-Na^+(aq) + \frac{1}{2}H_2(g)$

  or:

  $Na(s) + H^+(aq) \rightarrow Na^+(aq) + \frac{1}{2}H_2(g)$

- With **metal oxides** to form a salt and water:

  $CuO(s) + 2CH_3COOH(aq) \rightarrow (CH_3COO)_2Cu(aq) + H_2O(l)$

- With **alkalis** to form a salt and water:

  $NaOH(aq) + CH_3COOH(aq) \rightarrow CH_3COO^-Na^+(aq) + H_2O(l)$

- With **metal carbonates** to form a salt, water and carbon dioxide:

  $K_2CO_3(s) + 2CH_3COOH(aq) \rightarrow 2CH_3COO^-K^+(aq) + H_2O(l) + CO_2(g)$

> **Carboxylic acids** react with metal carbonates to form carbon dioxide gas. This reaction can be used to test for the presence of a carboxylic acid.

Reaction of carboxylic acids with alcohols to make esters

Carboxylic acids with react with alcohols, in the presence of **concentrated sulfuric(VI) acid** as catalyst, to form an equilibrium mixture in which the products are an ester and water.

Esters contain the functional group shown in Figure 23.11.

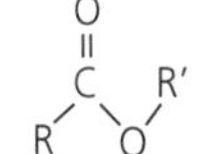

Figure 23.11 The ester functional group

In the example shown in Figure 23.12, methanoic acid is reacting with ethanol in the presence of concentrated sulfuric(VI) acid to form an ester called ethyl methanoate and water.

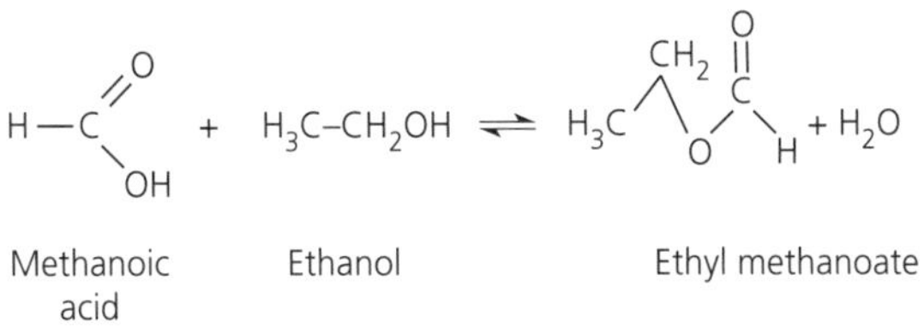

$HCOOH + H_3C{-}CH_2OH \rightleftharpoons HCOOCH_2CH_3 + H_2O$

Methanoic acid | Ethanol | Ethyl methanoate

Figure 23.12 An esterification reaction

In the esterification process, water is formed by combination of the H from the hydroxyl group of the alcohol and the O–H group from the carboxylic acid, as illustrated by the reaction of butanoic acid and methanol in Figure 23.13.

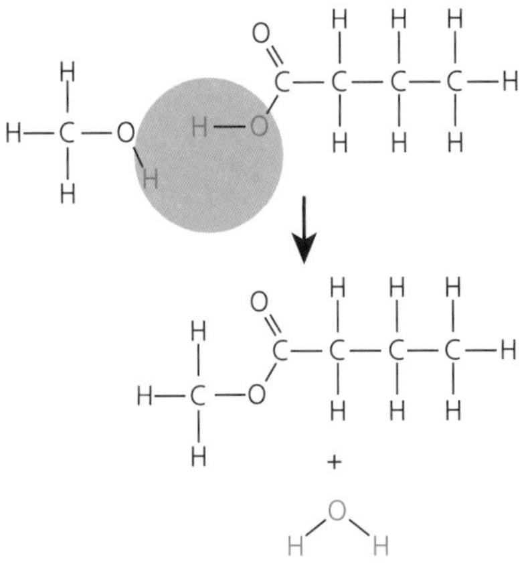

$CH_3OH + CH_3CH_2CH_2COOH \rightarrow CH_3CH_2CH_2COOCH_3 + H_2O$

Figure 23.13 The esterification process

Esters are used in the food and perfume industries as well as being good solvents and plasticisers.

Reactions of esters

REVISED

Hydrolysis reactions involving esters can take place in which either acids or bases may be used to catalyse the reactions. In these reactions, the acid or the base is a catalyst, so the processes are called **acid- or base-catalysed hydrolysis** reactions.

It is important to note that:

- in basic conditions, esters undergo complete hydrolysis, forming the corresponding alcohol and the sodium salt of the carboxylic acid
- in acidic conditions, esters are not completely hydrolysed — an equilibrium mixture is formed in which some ester is still present

Exam tip

When esters are hydrolysed in acid or basic solution, an alcohol is always formed but the carboxylic acid only forms in acidic conditions. A carboxylate anion forms in basic conditions.

Base-catalysed hydrolysis

Figure 23.14 shows the hydrolysis of methyl ethanoate in basic conditions as an example.

$$H_3C{-}O{-}C({=}O){-}CH_3 + OH^- \longrightarrow CH_3OH + {}^-O{-}C({=}O){-}CH_3$$

Methyl ethanoate — Methanol — Ethanoate ion

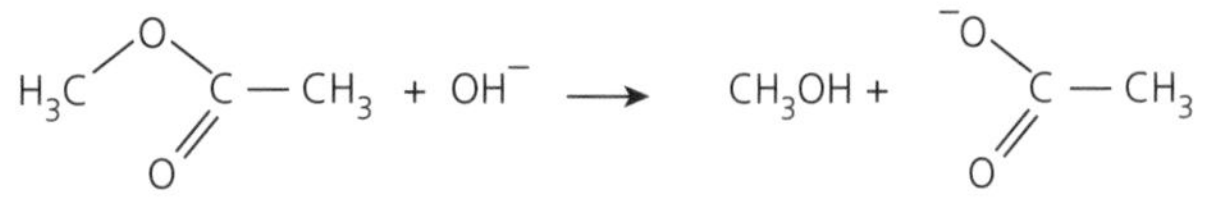

Figure 23.14 Base hydrolysis of an ester

The carboxylate anion, CH_3COO^-, can now react with water to reform the hydroxide ion catalyst.

Acid-catalysed hydrolysis

Figure 23.15 shows the hydrolysis of methyl ethanoate in acidic conditions as an example.

$$H_3C{-}O{-}C({=}O){-}CH_3 + H_2O \overset{H^+}{\rightleftharpoons} CH_3OH + HO{-}C({=}O){-}CH_3$$

Methyl ethanoate — Methanol — Ethanoic acid

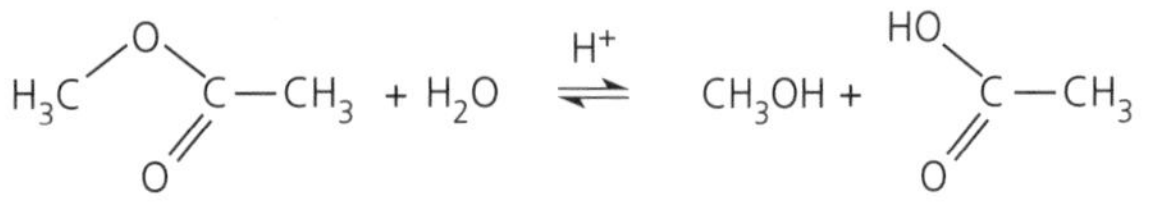

Figure 23.15 Acid hydrolysis of an ester

Naturally occurring fats and **oils** are esters made from propane-1,2, 3-triol (an alcohol commonly known as glycerol) and fatty acids (long-chain saturated or unsaturated carboxylic acids). If fats or oils are heated in concentrated **sodium hydroxide** solution (Figure 23.16), the ester is **hydrolysed** to form propane-1,2,3-triol (which is used in the food and cosmetics industries) and the sodium salts of the fatty acids (which are used in making soap).

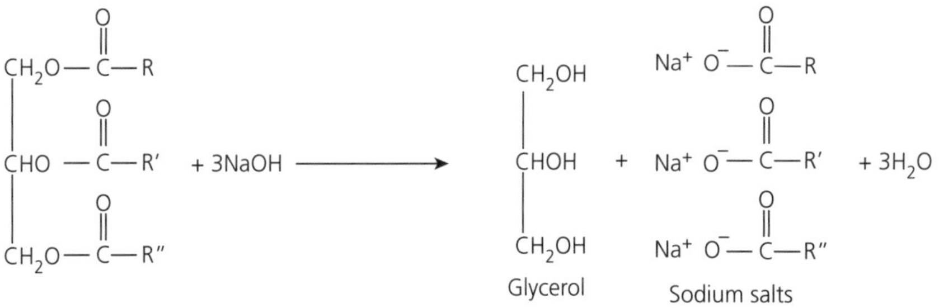

$$\begin{array}{l} CH_2O{-}C({=}O){-}R \\ CHO{-}C({=}O){-}R' \\ CH_2O{-}C({=}O){-}R'' \end{array} + 3NaOH \longrightarrow \begin{array}{l} CH_2OH \\ CHOH \\ CH_2OH \end{array} + \begin{array}{l} Na^+\ {}^-O{-}C({=}O){-}R \\ Na^+\ {}^-O{-}C({=}O){-}R' \\ Na^+\ {}^-O{-}C({=}O){-}R'' \end{array} + 3H_2O$$

Glycerol — Sodium salts of long-chain fatty acids

Figure 23.16 Soap making

Now test yourself

TESTED

4 Draw the structure of the products expected when the ester shown below is hydrolysed with (a) sodium hydroxide solution and (b) sulfuric(VI) acid.

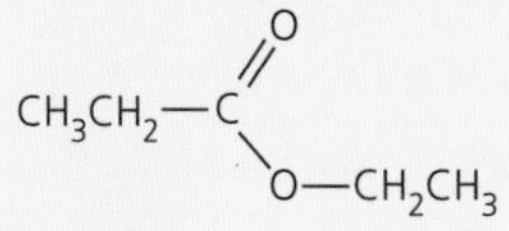

$CH_3CH_2-C(=O)-O-CH_2CH_3$

Answers on pp. 222–223

Revision activity

Test yourself on organic reactions by preparing sheets with three columns — 'Starting molecule', 'Reagent and conditions' and 'Product molecule'. Then cover up one of the columns and try to work it out from the remaining information in the other two columns. Do this every time you meet a new group of organic compounds.

Biodiesel

REVISED

When a naturally occurring fat or oil is heated with methanol (Figure 23.17) in the presence of a catalytic quantity of acid, glycerol (propane-1,2,3-triol) and methyl esters are formed. The methyl esters can be separated from the mixture and added to normal diesel to make **biodiesel** or used 'pure' as a fuel.

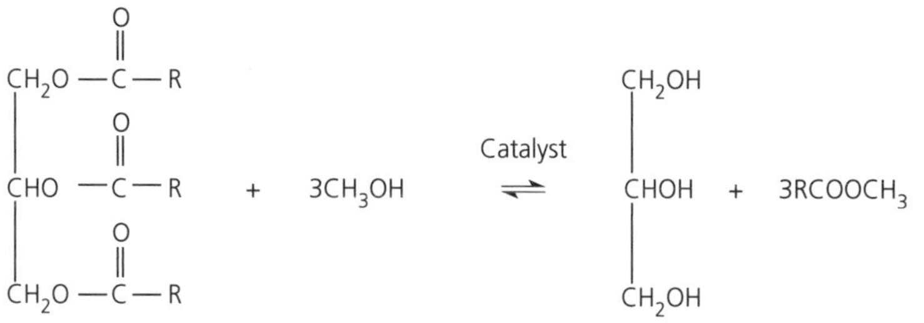

$$\begin{matrix} CH_2O-C(=O)-R \\ CHO-C(=O)-R \\ CH_2O-C(=O)-R \end{matrix} + 3CH_3OH \xrightleftharpoons{\text{Catalyst}} \begin{matrix} CH_2OH \\ CHOH \\ CH_2OH \end{matrix} + 3RCOOCH_3$$

Triglyceride — Methanol — Glycerol — Methyl esters

Figure 23.17 A source of biodiesel

Acylation

Acylation is the process of replacing a hydrogen atom in certain molecules by an RCO group, where R is an alkyl group. The structure of an acyl group is shown in Figure 23.18.

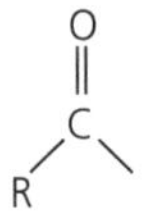

$R-C(=O)-$

Figure 23.18 The acyl functional group

Acylations can be carried out using acyl chlorides such as ethanoyl chloride, CH_3COCl, or by acid anhydrides such as ethanoic anhydride, $(CH_3CO)_2O$. Figure 23.19 shows the structures of these two reagents.

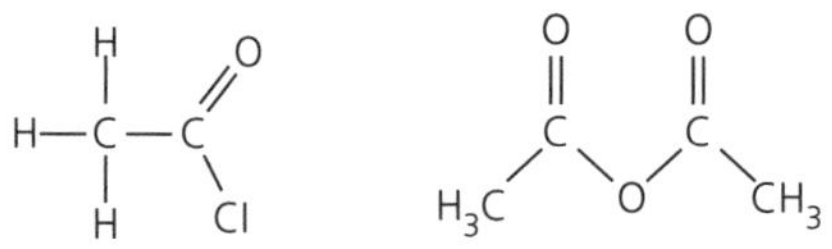

$H_3C-C(=O)-Cl$ $H_3C-C(=O)-O-C(=O)-CH_3$

Ethanoyl chloride — Ethanoic anhydride

Figure 23.19 Acylating agents

Reactions of ethanoyl chloride and ethanoic anhydride

REVISED

Nucleophiles such as water (H_2O), alcohols (ROH), ammonia (NH_3) and amines (RNH_2) all react with acyl chlorides and acid anhydrides in a predictable way.

Figure 23.20 shows the mechanism of the reaction between a nucleophile and an acyl chloride. This mechanism is called **nucleophilic addition–elimination**.

Intermediate

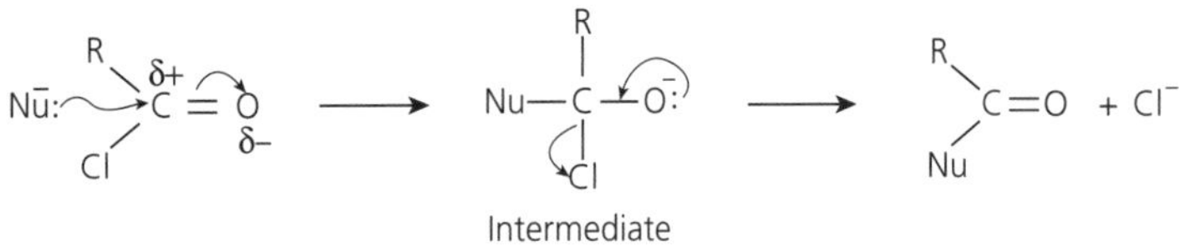

Figure 23.20 The mechanism of nucleophilic addition–elimination

The successive stages are:

1. The nucleophile donates a lone pair of electrons, making a bond with the electron-deficient carbon atom of the carbonyl group.
2. The high-energy π electrons in the C=O bond move to the oxygen atom to generate an intermediate anion.
3. A lone pair belonging to the oxygen atom then moves back into the C–O bond to regenerate the C=O bond.
4. The C–Cl bond breaks to release a chloride ion, Cl^-.

The overall effect is that the Cl group is replaced by a nucleophile.

When ethanoyl chloride reacts with an alcohol, the mechanism shown in Figure 23.21 occurs. The lone pair on the oxygen atom of the alcohol is used to bond the nucleophile to the carbon atom of the carbonyl group.

Revision activity

Learn these mechanisms and try to understand what is happening in each stage. In examinations, you may be asked to work through unfamiliar mechanisms in which you are provided with a general scheme.

On separate cards, draw a summary of each reaction mechanism you meet so that you will have a complete set at the end.

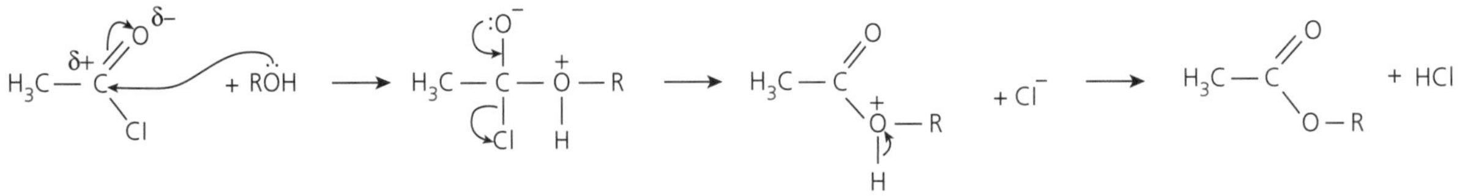

Figure 23.21 The acylation of an alcohol

When salicylic acid reacts with ethanoic anhydride (Figure 23.22), the oxygen atom of the hydroxyl group in salicylic acid donates a lone pair to one of the carbon atoms in a carbonyl group of the ethanoic anhydride. This triggers a rupturing of the ethanoic anhydride molecule so that the salicylate group can replace the ethanoate group.

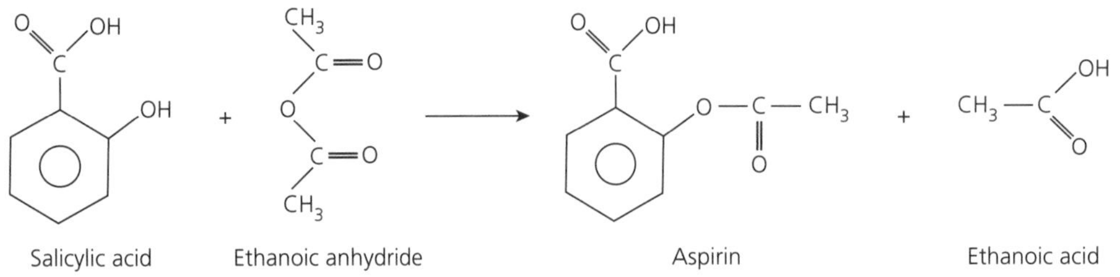

Salicylic acid | Ethanoic anhydride | Aspirin | Ethanoic acid

Figure 23.22 The synthesis of aspirin

When synthesising aspirin industrially, ethanoic anhydride is used instead of ethanoyl chloride because its reactions are more easily controlled. Ethanoyl chloride is difficult to store because it is easily hydrolysed by moisture and it releases corrosive HCl fumes.

Reactions of ethanoyl chloride

Ethanoyl chloride reacts with water, alcohols, ammonia and amines as follows.

- With water, H_2O:

  $CH_3COCl + H_2O \rightarrow CH_3COOH + HCl$

  making ethanoic acid — a carboxylic acid.
- With methanol, CH_3OH:

  $CH_3COCl + CH_3OH \rightarrow CH_3COOCH_3 + HCl$

  making methyl ethanoate — an ester.
- With ammonia, NH_3:

  $CH_3COCl + 2NH_3 \rightarrow CH_3CONH_2 + NH_4Cl$

  making ethanamide — an amide.
- With methylamine, CH_3NH_2:

  $CH_3COCl + CH_3NH_2 \rightarrow CH_3CONHCH_3 + HCl$

  making an N-substituted amide.

All this is shown schematically in Figure 23.23.

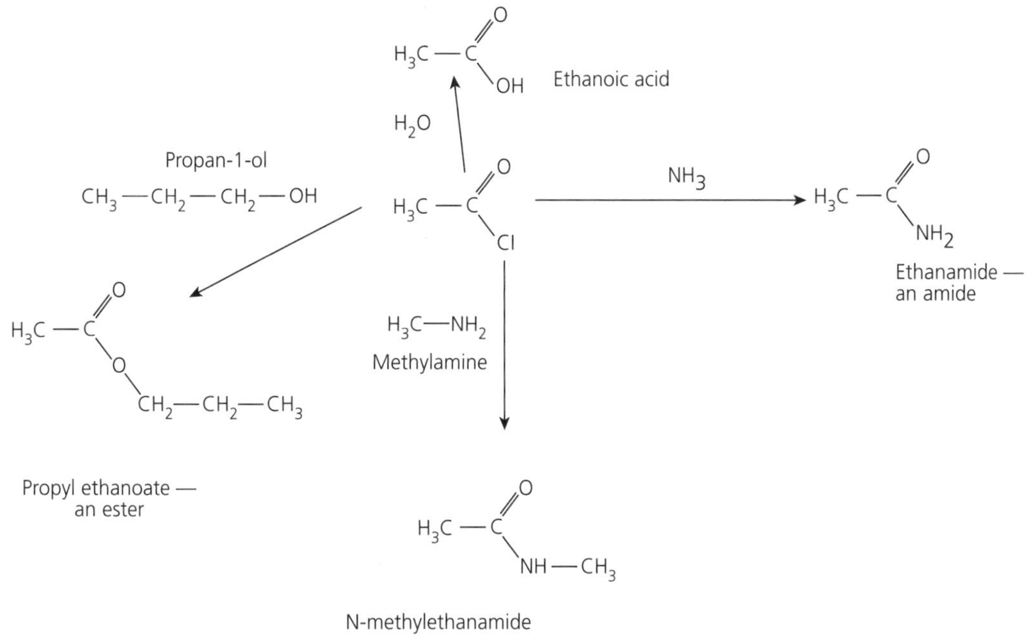

Figure 23.23 Reactions of ethanoyl chloride

Reactions of ethanoic anhydride

Ethanoic anhydride reacts with water, alcohols, ammonia and amines as follows. Notice how the main organic product is the same as that formed by ethanoyl chloride. The only difference is that CH_3COOH is formed rather than HCl.

- With water, H_2O:

  $(CH_3CO)_2O + H_2O \rightarrow 2CH_3COOH$

  making ethanoic acid — a carboxylic acid.
- With methanol, CH_3OH:

  $(CH_3CO)_2O + CH_3OH \rightarrow CH_3COOCH_3 + CH_3COOH$

  making methyl ethanoate — an ester.

- With ammonia, NH_3:

$(CH_3CO)_2O + 2NH_3 \rightarrow CH_3CONH_2 + CH_3COONH_4$

making ethanamide — an amide.

- With methylamine, CH_3NH_2:

$(CH_3CO)_2O + CH_3NH_2 \rightarrow CH_3CONHCH_3 + CH_3COOH$

making an N-substituted amide.

Now test yourself

TESTED

5 Draw the structure of the organic product formed when the compound below reacts with ethanoic anhydride, $(CH_3CO)_2O$.

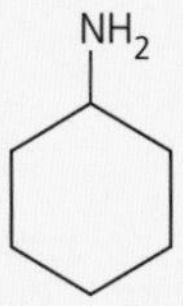

NH_2

Answers on p. 223

Exam practice

1 (a) Give the molecular formula of propanal. [1]

(b) Draw the structure of the organic product formed when propanal reacts with:

(i) $NaBH_4$(aq) [1]

(ii) $AgNO_3$(aq) and NH_3(aq) (Tollens' reagent) [1]

(iii) HCN with a trace of KCN [1]

(c) The product from the reaction in part (b)(iii) is chiral. Draw the enantiomeric forms of the product. [2]

2 Compounds A–C have the following molecular formulae:

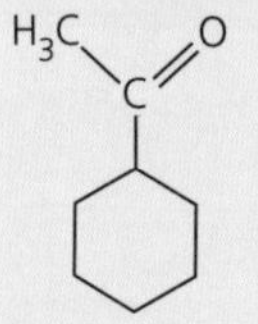

A 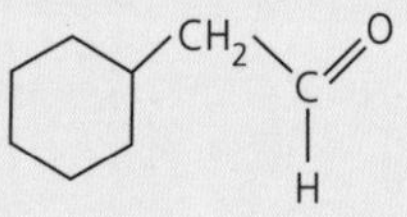 B 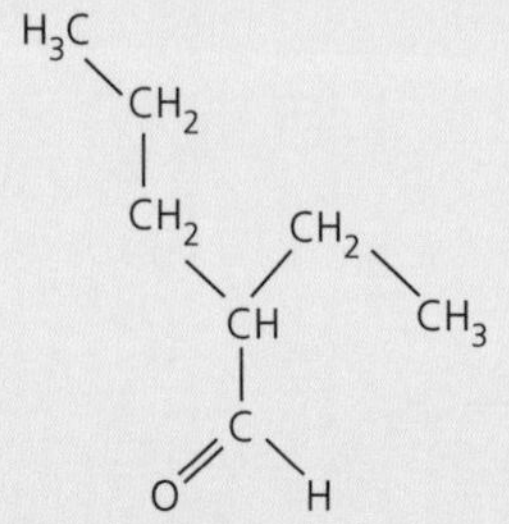 C

(a) Describe a chemical test that would enable you to distinguish between compounds A and B. [2]

(b) Write an equation, using [H], to show how compound B reacts with $NaBH_4$. [2]

(c) Draw the structure of the organic compound formed when C reacts with Fehling's solution. [1]

3 Give the reagents and conditions required for the chemical transformations:

(a) Y to X [2]

(b) Y to Z [2]

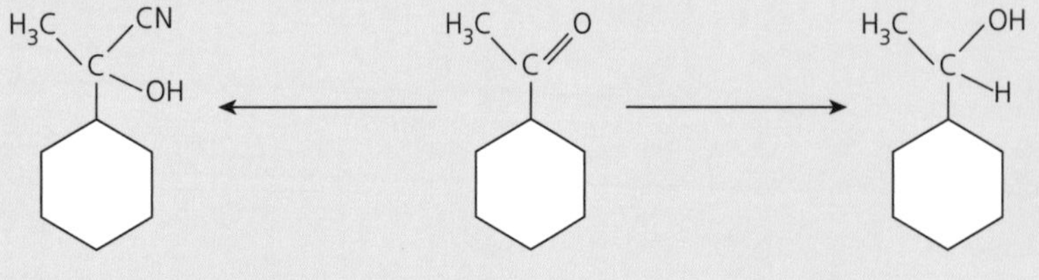

X Y Z

Answers and quick quiz 23 online

ONLINE

Summary

You should now have an understanding of:

- the functional groups in aldehydes and ketones
- the carbonyl group and how it is polarised
- how to distinguish between aldehydes and ketones
- the reactions of the carbonyl group
- the mechanism of nucleophilic addition
- carboxylic acids and how they react as acids
- how carboxylic acids react with alcohols in the presence of an acid catalyst to form an ester
- how esters can be hydrolysed either in acidic or basic conditions
- what is meant by 'biodiesel'
- the acylation reaction
- the reactions of acyl chlorides and acid anhydrides as typical acylating agents
- the mechanism by which acyl chlorides react — the nucleophilic addition–elimination mechanism

24 Aromatic compounds

Structure of benzene

Benzene is an unsaturated cyclic hydrocarbon with molecular formula C_6H_6. The molecule has a planar hexagonal structure. It consists of a σ–bonded framework in which all the H–C–H angles are 120°. There is also a delocalised π–electron system above and below the plane of atoms. The π system is formed by the overlap of p orbitals on adjacent carbon atoms, as shown in Figure 24.1.

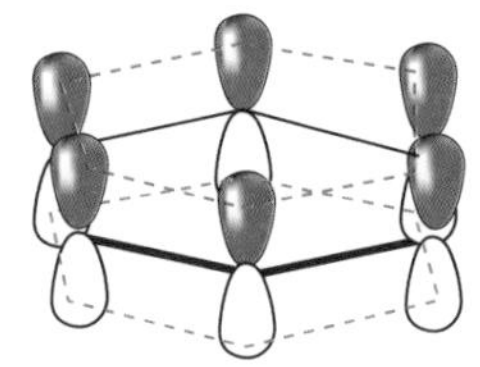

Figure 24.1 Delocalised π electrons above and below the plane of the benzene ring

- The π system above and below the benzene ring consists of six π electrons that are delocalised. This enhances the thermodynamic stability of the ring.
- The carbon–carbon bonds in benzene are intermediate in length between those of a single bond and a double bond (C–C = 0.154 nm; C=C = 0.134 nm; benzene = 0.139 nm). This is evidence of delocalisation in that intermediate-length bonds are formed.
- The internal bond angles in the benzene ring are 120°. All six carbon atoms have a trigonal planar arrangement — three σ bonds are formed by each. with the fourth electron of each carbon in the delocalised π cloud.

Exam tip

When discussing the structure of benzene, refer to the overlapping of carbon atom p orbitals to form the σ bonds to hydrogen atoms and the delocalised π cloud of electrons.

The benzene ring is best represented as shown in Figure 24.2.

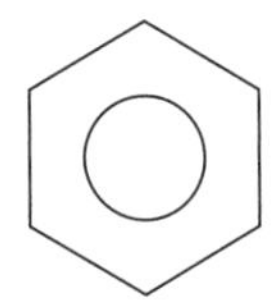

Figure 24.2 Representation of the benzene ring

However, it is sometimes convenient to consider the Kekulé form shown in Figure 24.3.

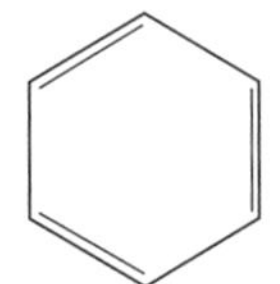

Figure 24.3 The Kekulé form for benzene

The Kekulé form is not strictly correct because it implies that separate double and single bonds exist in the molecules. We now know this not to be the case and that delocalisation takes place (see the evidence above).

Notice that the hydrogen atoms are not normally indicated in these structures.

Benzene stability

REVISED

Benzene is more thermodynamically stable than expected. The lowering of its energy due to the delocalisation can be quantified by using a simple energy level diagram (Figure 24.4).

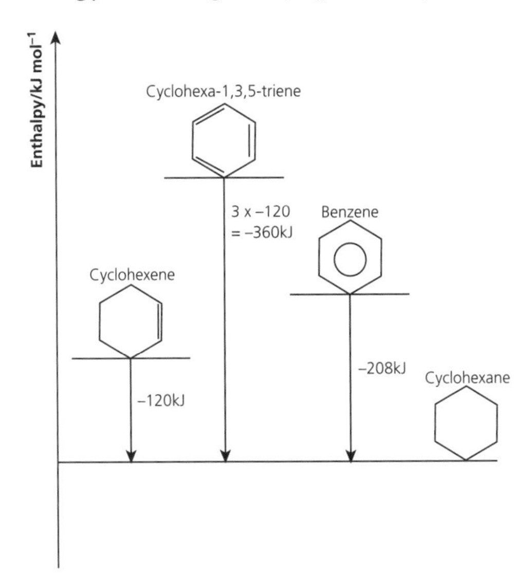

Figure 24.4 Hydrogenation enthalpy evidence for benzene's structure

In all the reactions in Figure 24.4, a process of hydrogenation is represented — a simple addition reaction of hydrogen with three substances:

- 'Real' benzene, in which delocalisation is assumed. The hydrogenation enthalpy is $-208\,kJ\,mol^{-1}$:

  $C_6H_6(g) + 3H_2(g) \rightarrow C_6H_{12}(g)$

- Cyclohexene — a cyclic alkene that has only one carbon–carbon double bond. Its enthalpy of hydrogenation is $-120\,kJ\,mol^{-1}$:

  $C_6H_{10}(g) + H_2(g) \rightarrow C_6H_{12}(g)$

- Cyclohexa-1,3,5-triene — a 'theoretical' compound that has three carbon–carbon double bonds and no delocalisation is assumed. The hydrogenation enthalpy is estimated to be three times the value for cyclohexene, i.e. $-120 \times 3 = -360\,kJ\,mol^{-1}$:

  $C_6H_6(g) + 3H_2(g) \rightarrow C_6H_{12}(g)$

It can be seen from Figure 24.4 that the 'real' benzene is $152\,kJ\,mol^{-1}$ lower in energy than the 'theoretical' benzene molecule. This lowering in energy, due to the delocalised electrons, explains why the reactions of benzene have higher activation energies than expected. They are also slower and require higher temperatures and more 'severe' conditions than the reactions of a typical alkene.

Exam tip

Try to remember the evidence for delocalisation in benzene — data regarding bond lengths, bond angles and enthalpy of hydrogenation are all useful to know.

Reactions of benzene

Most of the reactions of benzene are **electrophilic substitution** processes. In these reactions, one or more of the hydrogen atoms in benzene is/are substituted by other group(s).

Nitration of the benzene ring

REVISED

Nitrating a benzene ring involves placing a nitro group ($-NO_2$) into the benzene ring instead of a hydrogen — the process is called **nitration**. Nitrobenzene has the structure shown in Figure 24.5.

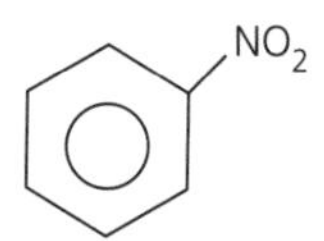

Figure 24.5 Nitrobenzene

An equation for the reaction is shown in Figure 24.6.

NO_2 + HNO_3 $\xrightarrow[50°C]{H_2SO_4}$ + H_2O

Figure 24.6 Nitration of benzene

> **Exam tip**
>
> The reagents for carrying out nitration are concentrated nitric(v) acid, $HNO_3(l)$, a catalytic amount of concentrated sulfuric(vi) acid, $H_2SO_4(l)$, and a maximum temperature of 50°C.

Electrophilic substitution mechanism

An electrophile called the nitronium ion, NO_2^+, is produced:

$$HNO_3 + H_2SO_4 \rightleftharpoons HSO_4^- + H_2NO_3^+$$

$$H_2NO_3^+ \rightarrow NO_2^+ + H_2O$$

The electrophile then reacts with the benzene ring (Figure 24.7).

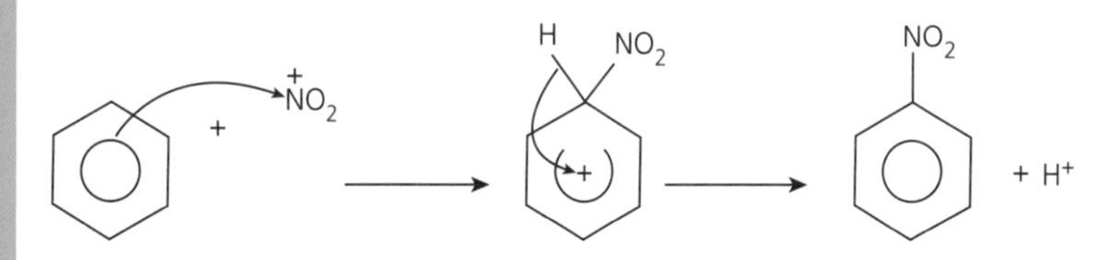

Figure 24.7 Electrophilic substitution — nitration

> **Exam tip**
>
> It is important to know the electrophilic substitution mechanism — there are several variations of it that are very similar. Remember and know one of them, and the others can often be worked out.

- Two electrons are donated from the π cloud of delocalised electrons in the benzene ring.
- A new carbon–nitrogen bond forms in an intermediate species. The benzene ring in this has only partial delocalisation — this raises its energy.
- A carbon–hydrogen bond breaks and the two electrons in the bond move back into the π system on the benzene ring to restore the delocalisation.
- The resulting proton, H^+, recombines with the hydrogensulfate(vi) ion, HSO_4^-, to regenerate the H_2SO_4 catalyst.

Uses of nitration

Many of the nitro products that can be formed by this reaction are used for making explosives — for example, 2-methyl-1,3,5-trinitrobenzene (Figure 24.8), otherwise known as 2,4,6-trinitrotoluene (TNT).

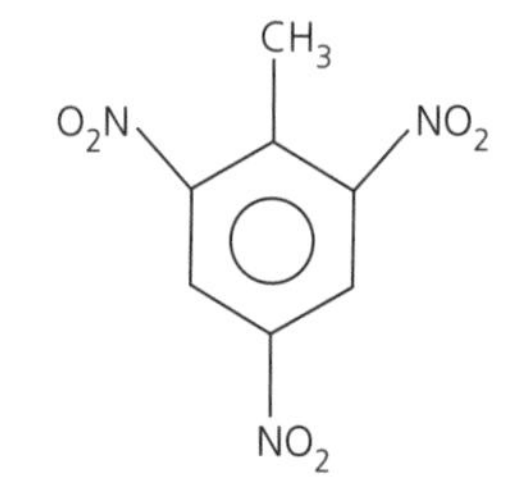

Figure 24.8 TNT

Reduction of these aromatic nitro compounds is achieved by refluxing with tin in concentrated hydrochloric acid — this yields the corresponding aromatic amine.

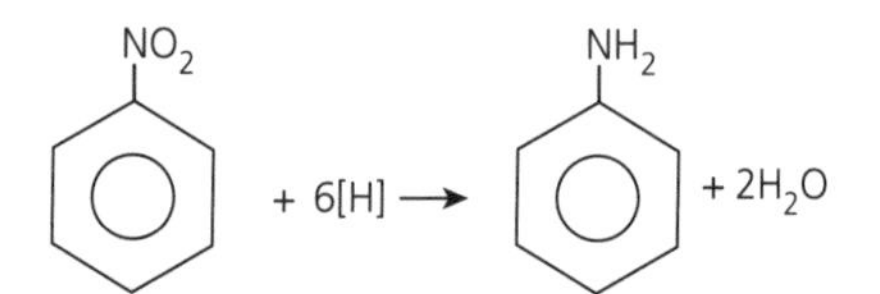

Figure 24.9 Making phenylamine

The phenylamine formed using this reaction (Figure 24.9) can be used for making **dyestuffs**.

Friedel–Crafts acylation

REVISED

Acylation involves incorporating an RCO group into a molecule. Figure 24.10 shows the reaction between benzene and ethanoyl chloride in the presence of aluminium chloride, $AlCl_3$.

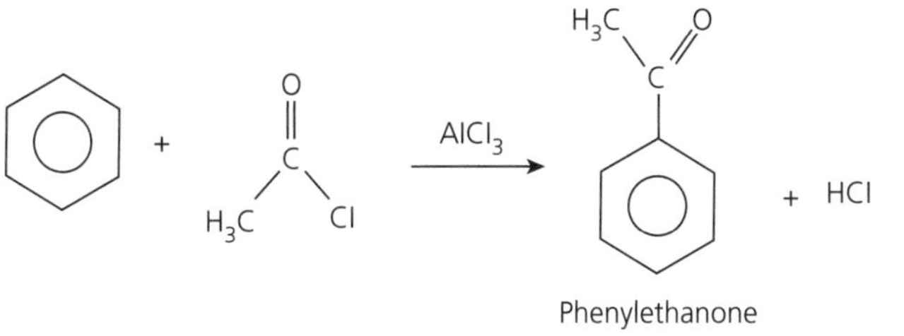

Phenylethanone

Figure 24.10 Making phenylethanone

The mechanism for the reaction is **electrophilic substitution**.

The required electrophile (the acylium ion), CH_3C^+O, is formed using an aluminium chloride catalyst, $AlCl_3$ (Figure 24.11).

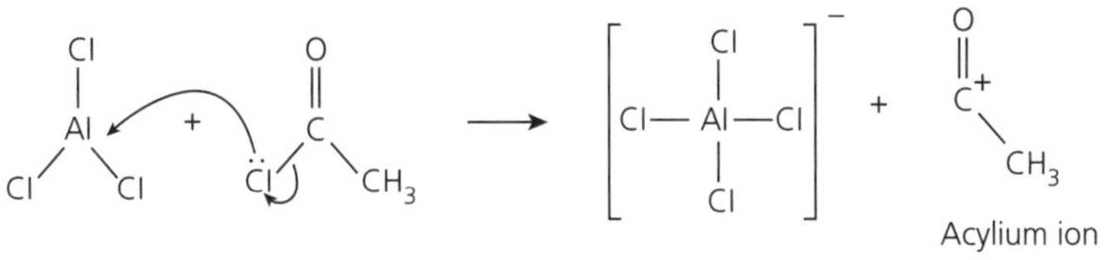

Acylium ion

Figure 24.11 Making the electrophile

Two π electrons from the benzene ring are donated to the positive carbon of the acylium ion and a new C–C bond is formed (Figure 24.12). Loss of a proton from the intermediate then regenerates the delocalised π cloud in the benzene ring.

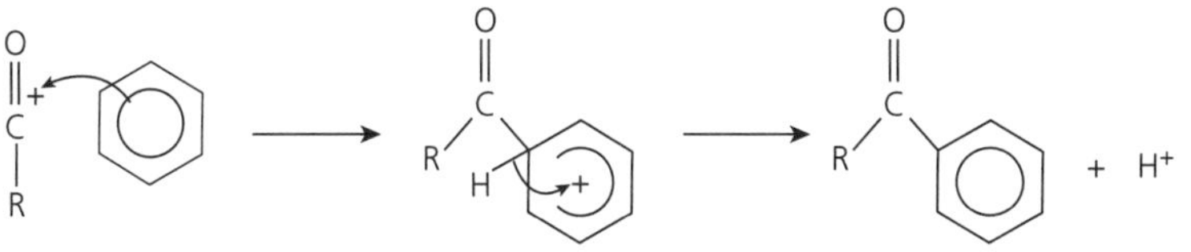

Figure 24.12 The substitution step

The released proton can react with the $AlCl_4^-$ ion to regenerate the catalyst:

$$AlCl_4^- + H^+ \rightarrow AlCl_3 + HCl$$

Exam tip

This mechanism is virtually the same as the nitration mechanism covered earlier, except that the electrophile is CH_3C^+O instead of N^+O_2.

Revision activity

As the number of reactions you need to know starts to build up, produce spider diagrams or reaction schemes (in colour) and keep revising them.

Now test yourself

TESTED

1 Examine the reactions below.

Reaction 1:

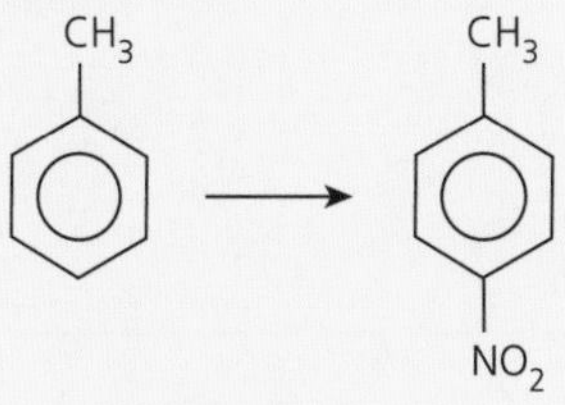

Reaction 2:

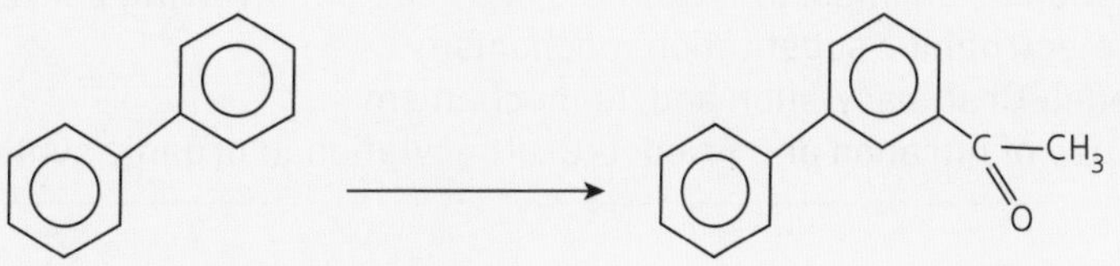

(a) Give the formula of the electrophile featured in each reaction.
(b) Give the reagents and conditions required for each reaction.

Answer on p. 223

Exam practice

1 The questions below relate to this reaction scheme:

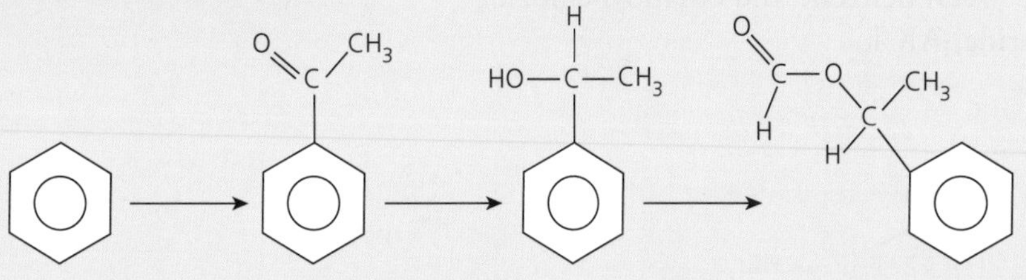

A → B → C → D

(a) Give the reagents required for each stage in the reaction sequence. [3]
(b) Explain why a racemic mixture forms when C is synthesised. [2]
(c) Suggest the name of a reagent that could be used to convert C back into B. [1]
(d) Name and outline the mechanism for the formation of B from A. [3]

2 The following scheme enables compound M to be synthesised from benzene.

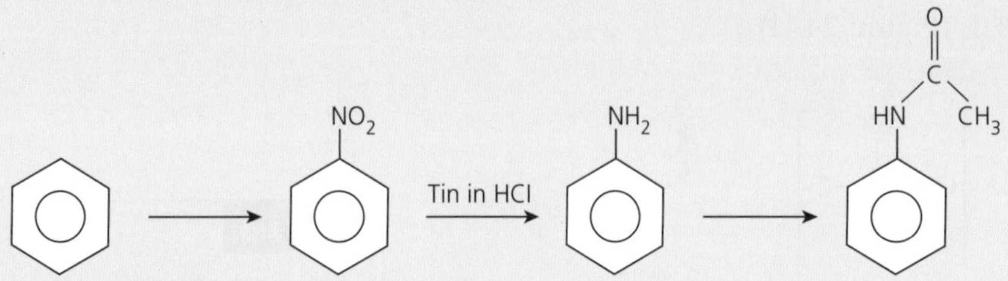

J → K (NO$_2$) —Tin in HCl→ L (NH$_2$) → M

(a) What is the molecular formula for compound M? [1]
(b) What type of reaction takes place when K is converted to L? [1]
(c) Suggest reagents and conditions required for the conversion from:
  (i) J to K [1]
  (ii) L to M [1]
(d) Write an equation for the transformation J into K. [1]
(e) Give the names of the mechanisms involved in the conversion from:
  (i) J to K [1]
  (ii) L to M [1]
(f) Suggest a use for compounds that can be formed using the type of reaction when J is converted into K. [1]

Answers and quick quiz 24 online

ONLINE

Summary

You should now have an understanding of:

- the structure of benzene in terms of the π cloud of delocalised electrons and how this is formed by atomic orbital overlap
- the evidence to support the idea of electron delocalisation in benzene rings
- nitration as an important example of a reaction involving benzene
- the electrophilic substitution mechanism
- Friedel–Crafts acylation and its mechanism
- the use of nitration and Friedel–Craft acylation in organic syntheses

25 Amines and amino acids

Amines can be thought of as alkylated ammonia species in which ammonia's hydrogen atoms are substituted by one or more alkyl groups.

Amine molecules, like ammonia, are pyramidal in their spatial arrangement of covalent bonds around the nitrogen atom (Figure 25.1), with the internal angle being approximately 107°.

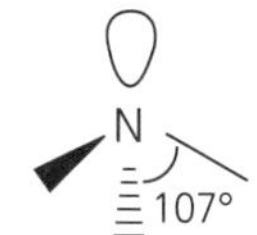

Figure 25.1 Spatial arrangement around a saturated nitrogen atom

Amines are classed according to the number of alkyl groups (or C–N) bonds surrounding the central nitrogen atom.

- Primary amines have one attached alkyl group:

  H, N, C_2H_5, H

  Ethylamine
- Secondary amines have two attached alkyl groups:

  H, N, H_3C, CH_3

  Dimethylamine
- Tertiary amines have three attached alkyl groups:

  N

  Triphenylamine
- Quaternary ammonium salts have four attached alkyl groups:

  $[N(CH_3)_4]^+ Cl^-$

  Tetramethylammonium chloride

Amines as bases

Phenylamine — an aromatic amine

REVISED

Phenylamine (Figure 25.2) consists of an amino group ($-NH_2$) attached directly to a benzene ring.

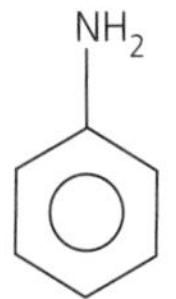

Figure 25.2 Phenylamine

Nitrogen's lone pair of electrons can overlap with the π-system in the benzene ring and, as a result, the lone pair of the nitrogen atom is delocalised into the π-system.

Phenylamine is a weaker Brönsted–Lowry base than ammonia because the nitrogen atom's lone pair is delocalised into the π-system and is, therefore, less available for protonation:

$$C_6H_5NH_2 + H^+ \rightarrow C_6H_5NH_3^+$$

Aliphatic amines in which the alkyl groups are attached directly to the nitrogen atom have a slightly enhanced electron density on the nitrogen atom, due to the electron-donating properties of the alkyl groups. This makes the nitrogen lone pair easier to protonate than in ammonia (no alkyl groups) and much easier than in phenylamine.

Aliphatic amines generally increase in base strength as the number of alkyl groups attached to the nitrogen atom increases. Each alkyl group donates electrons towards the nitrogen atom, so the more alkyl groups there are, the greater will be the electron density on the nitrogen, and the easier protonation will take place.

The order, with strongest base first, is:

aliphatic amines > ammonia > phenylamine

Exam tip

When explaining the relative basic strength of amines, use the correct terms in your explanation — Brönsted–Lowry base, lone pair, delocalised, protonation, π-system, electron density etc.

Amines as nucleophiles

Amines, like ammonia, are nucleophiles and so react with haloalkanes (by nucleophilic substitution) — see page 82 — and with acyl chlorides (by nucleophilic addition–elimination) — see page 185.

Amines reacting with haloalkanes

REVISED

Haloalkanes have a polar carbon–halogen bond ($^{\delta+}C–X^{\delta-}$). This is attacked by the nitrogen atom in amine molecules, displacing the halogen as a halide ion.

An excess of ammonia (in ethanol solvent) can react with bromoethane, C_2H_5Br:

$$C_2H_5Br + NH_3 \rightarrow C_2H_5NH_2 + HBr$$

but preferably:

$$C_2H_5Br + 2NH_3 \rightarrow C_2H_5NH_2 + NH_4Br$$

This is because the acidic HBr and any remaining ammonia in the reaction mixture react immediately to form ammonium bromide.

Iodomethane reacts with ethylamine (a primary amine):

$$CH_3I + C_2H_5NH_2 \rightarrow C_2H_5NHCH_3 + HI$$

This forms ethylmethylamine — a secondary amine. Depending on conditions the protonated amine iodide salt $[C_2H_5NH_2CH_3]^+I^-$ may form.

If excess iodomethane is present this process can continue, producing a mixture of products (Figure 25.3). Secondary amines, tertiary amines and quaternary ammonium salts may all be present depending on the amount of iodomethane present.

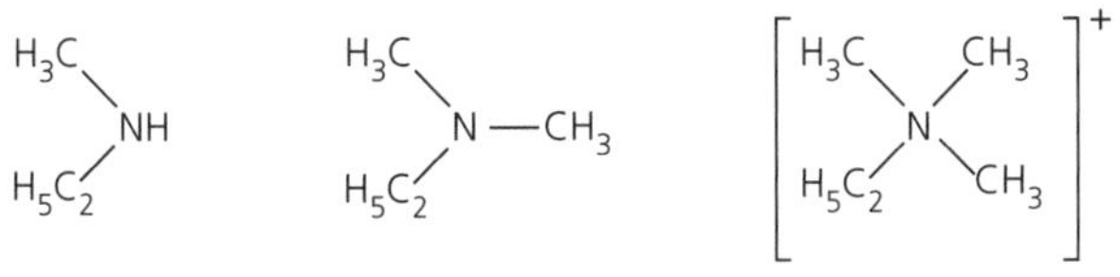

Secondary amine

Tertiary amine

Quaternary ammonium ion

Figure 25.3 Successive reaction products

Quaternary ammonium salts are used as **cationic surfactants** in detergents. Examples of two of these are shown in Figure 25.4. The polar/charged nature (hydrophilic) of one end of the ion and the hydrophobic nature of the hydrocarbon chain can be exploited in their action as detergents, particularly in removing unwanted grease.

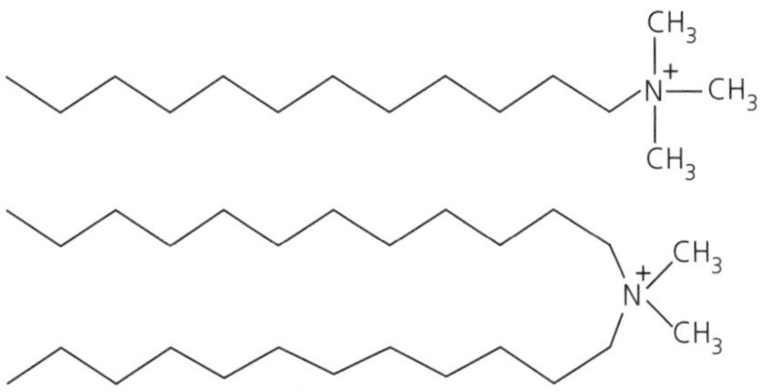

Figure 25.4 Detergents

Nucleophilic substitution mechanism

In the reaction between a bromoalkane and ammonia (Figure 25.5):

- the ammonia molecule acts as a **nucleophile** by donating its lone pair of electrons to the slightly positively charged carbon atom in the $^{\delta+}C{-}Br^{\delta-}$ bond.
- A new carbon–nitrogen bond forms.
- An ammonia molecule (acting as a base) then removes a proton from the $-NH_3^+$ group, forming an amine.

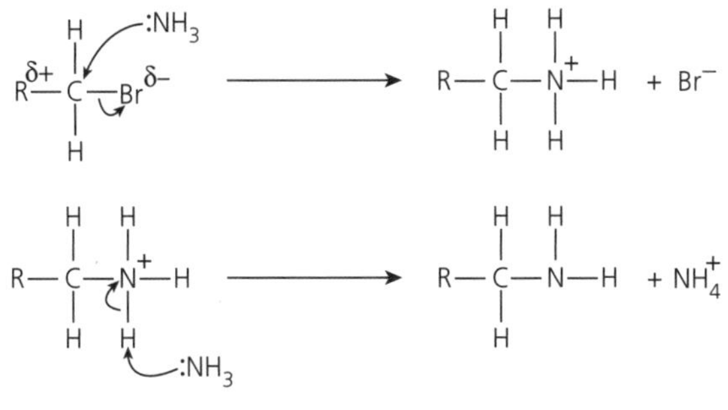

Figure 25.5 Nucleophilic substitution

Exam tip

This mechanism is similar to the AS mechanism in which hydroxide ions, OH^-, act as nucleophiles when attacking a halogenoalkane. The difference is the need for a second ammonia molecule to remove a proton in the amine reaction.

Exam tip

The rate of this reaction depends on the strength of the C–X bond — iodoalkanes react faster than bromoalkanes because a C–I bond is weaker than a C–Br bond.

Synthesis of amines

Aliphatic amines

REVISED

Nitriles are organic molecules that contain a CN group. They can be reduced to primary amines using hydrogenation, with the help of a nickel catalyst:

$$CH_3CN + 2H_2 \rightarrow CH_3CH_2NH_2$$

Ethylamine

Nitriles can also be reduced using **sodium tetrahydridoborate(III), $NaBH_4$**, in a non-aqueous solvent, like ethanol, to form primary amines:

$$C_2H_5CN + 4[H] \rightarrow C_2H_5CH_2NH_2$$

Propylamine

This reaction is also reduction. A balanced equation can be written using [H] to represent the role of the $NaBH_4$.

Aromatic amines

REVISED

Aromatic amines such as phenylamine (Figure 25.6) can be prepared by reducing the appropriate aromatic nitro compound by refluxing with tin in concentrated hydrochloric acid.

NO_2 (benzene ring) + 6[H] —(tin in HCl, reflux)→ NH_2 (benzene ring) + $2H_2O$

Nitrobenzene   Phenylamine

Figure 25.6 Preparation of phenylamine

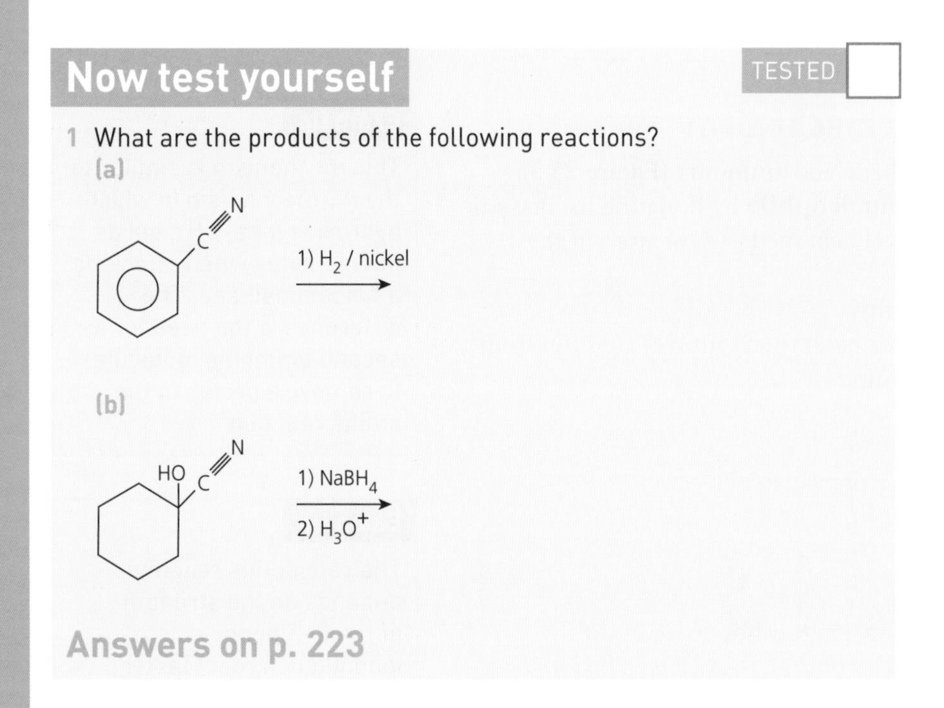

Amino acids

The general formula of an α-amino acid is $H_2NC(R)(H)COOH$. Figure 25.7 shows its structure.

- The α-carbon atom is the one (*) that is linked to both the COOH and NH_2 group — all naturally occurring amino acids have this feature.
- The R group can vary, but when it is a hydrogen atom, as in glycine, the molecule is not chiral because two of the four groups attached to the central carbon atom are identical. However, all other amino acids are chiral and can exist in either the (+) or (−) enantiomeric forms. It is an interesting feature of their stereochemistry that all optically active amino acids are of the 'left-hand' variety or (−) form. This means that they all rotate the plane of polarised light to the left.

R
$H_2N—C^*—COOH$
H

Figure 25.7 General formula for an α-amino acid

- The enantiomeric forms of a general amino acid are shown in Figure 25.8 — note their mirror image relationship to each other.

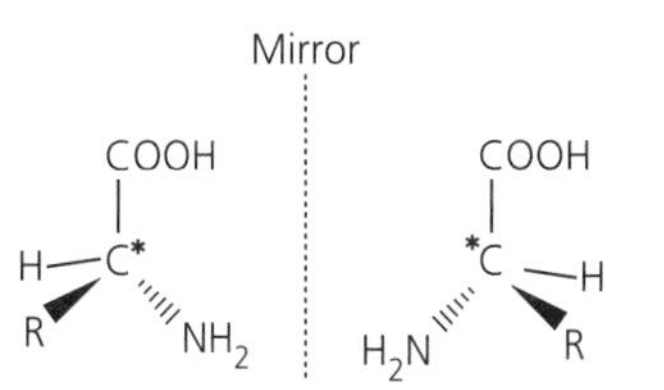

Figure 25.8 Chirality in α-amino acids

Alanine and serine (Figure 25.9) are further examples of naturally occurring amino acids — both are chiral (unlike glycine).

CH_3 / $H_2N—C—COOH$ / H — Alanine

CH_2OH / $H_2N—C—COOH$ / H — Serine

Figure 25.9 Two simple α-amino acids

Acid–base reactions of amino acids

REVISED

Amino acids have both the basic amine group (NH_2) and the acidic carboxylic acid group (COOH). When an acid or a base is added to an amino acid, the base or acid group reacts, as illustrated by alanine in Figure 25.10.

$H_2N—C(CH_3)(H)—COOH$

OH^- → $H_2N—C(CH_3)(H)—COO^-$

H^+ → $H_3\overset{+}{N}—C(CH_3)(H)—COOH$

Figure 25.10 Alanine acting as a base and an acid

> **Exam tip**
>
> Learn these reactions — they are easy to understand. The NH_2 group is protonated in acids, whereas the COOH group loses a proton in a basic environment.

An **isoelectronic point** is the pH at which a **zwitterion** ('double' ion) of an amino acid exists — the overall charge being zero. At an isoelectronic point both the NH_3^+ and COO^- groups are present. Solid amino acids have a higher than expected melting point because of the presence of zwitterions.

Proteins

Amino acids are able to combine with each other, in a condensation process, to form **proteins** (Figure 25.11). Amino acids are joined to each other by **peptide links**.

$H_2N—C(R')(H)—C(=O)OH + H_2N—C(R'')(H)—C(=O)OH \longrightarrow H_2N—C(R')(H)—C(=O)—N(H)—C(R'')(H)—C(=O)OH + H_2O$

Peptide link

Figure 25.11 Protein formation

The short protein strand shown in Figure 25.11 consists of two amino acid 'residues' and is called a **dipeptide**.

It is possible to **hydrolyse** a protein, either in acidic or basic conditions, to form the constituent amino acids. Figure 25.12 shows the hydrolysis of a tripeptide — a short protein strand consisting of three amino acid 'residues' — to form three amino acids.

Peptide links

$H_2N-C(R')(H)-C(=O)-N(H)-C(R'')(H)-C(=O)-N(H)-C(R''')(H)-COOH$

Hydrolysis

Amine group

Carboxyl group

$H_2N-C(R')(H)-COOH + H_2N-C(R'')(H)-COOH + H_2N-C(R''')(H)-COOH$

Amino acids (side chains are R′, R″ and R‴)

Figure 25.12 Hydrolysis of a protein

The amino acids formed can be separated by **chromatography**. Analysis of a chromatogram can reveal the structure of the original, although it is a complex procedure.

Hydrogen bonding in proteins

REVISED

Protein chains can interact with other protein chains using hydrogen bonds. Figure 25.13 shows the interaction between a carbonyl group, $^{\delta+}C=O^{\delta-}$, in one peptide link with the $^{\delta+}N-H^{\delta+}$ group of another peptide link in another protein strand.

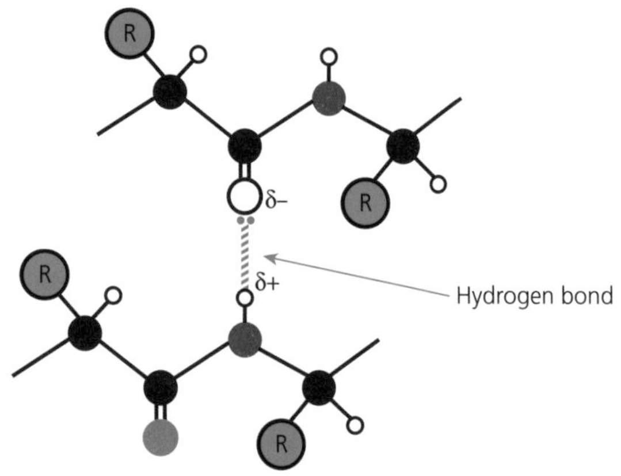

Figure 25.13 Hydrogen bonding in proteins

β-pleated sheets and α-helices are two of several 3-dimensional forms of proteins that owe their structures to the existence of hydrogen bonds.

Primary, secondary and tertiary structures of proteins

REVISED

There are various ways in which we may want to study the structure of a protein (Figure 25.14).

- The **primary structure** describes the sequencing or ordering of the amino acids along the protein strand.
- The **secondary structure** refers to the interaction between protein strands, using hydrogen bonding to form sheets and various other shapes in which proteins can fold and contort.
- The **tertiary structure** describes how the secondary structures can interact with other protein structures to form more complex species.

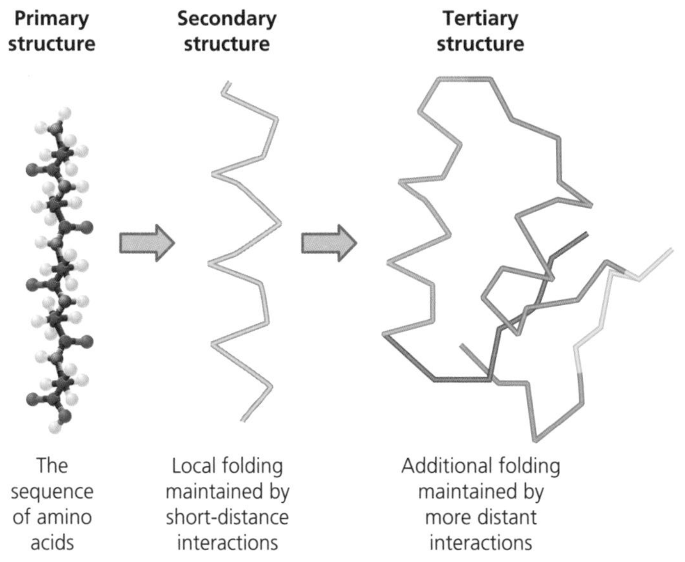

Figure 25.14 Protein structure

Now test yourself

TESTED

2 The two amino acids shown below are commonly called alanine and aspartic acid.

$H_3C-CH(NH_2)-COOH$

Alanine

$HOOC-CH_2-CH(NH_2)-COOH$

Aspartic acid

Draw the structures of:

(a) the zwitterion of alanine
(b) the alanine species that would exist at pH 3
(c) three dipeptides that could be formed by alanine and aspartic acid

Answers on p. 223

Enzymes

REVISED

Enzymes are complex proteins.

The action of enzymes as catalysts involves a **stereospecific** (or **chiral) active site** that binds to a substrate molecule. The stereospecific active site will only 'recognise' another chiral molecule of the correct chirality. This molecule can then bind effectively to the active site because the functional groups of the substrate and the active site can align themselves correctly. The other molecule of the incorrect chirality is not affected by the enzyme.

Mixtures containing equal amounts of both chiral forms of a molecule — a **racemic mixture** — can be separated by using an enzyme that has a chiral active site. One of the molecules can then bind with the active site, and therefore react, producing new product molecules, whereas the other molecule is left intact and can then be separated from the mixture. This process is called **resolution**.

The principle of many drug actions involves binding to the active site of an enzyme, thereby blocking the active site, so that the enzyme is not affected by other target molecules. Computers can be used to help design such drugs. If the structure of the active site of the enzyme is understood, molecules can be 'engineered' by computer modelling.

DNA

The structure of a DNA molecule consists of a double helix, which maintains its structure using paired bases holding the strands together like a 'zip' (Figure 25.15).

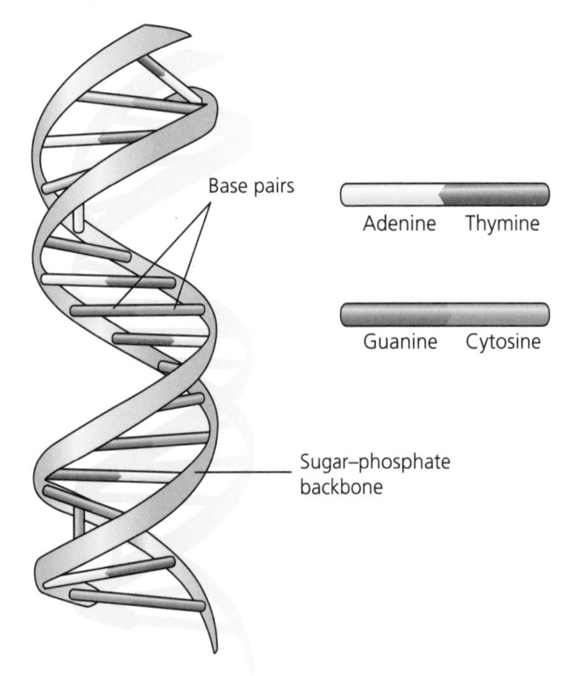

Figure 25.15 DNA structure

The backbone of each DNA strand consists of many **nucleotides** bonded together. A nucleotide consists of a deoxyribose sugar, a phosphate group and a base all bonded together covalently.

Deoxyribose (Figure 25.16) is based on a simple sugar molecule (ribose).

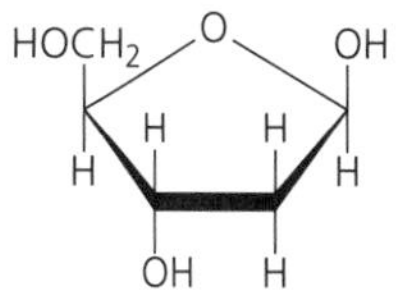

Figure 25.16 Deoxyribose

Figure 25.17 shows the typical structure of a nucleotide, in which the three main units are chemically bonded together (phosphate, ribose and base).

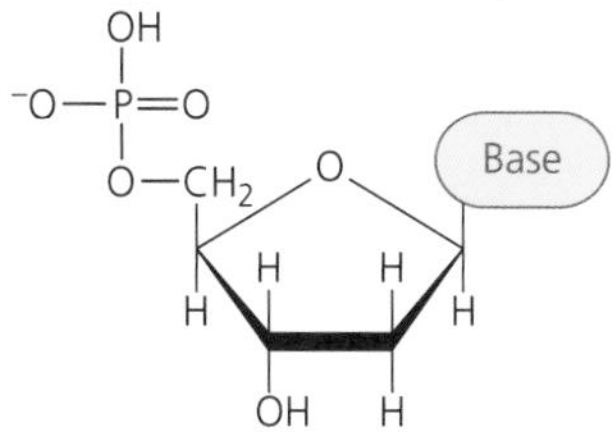

Figure 25.17 Structure of a nucleotide

Nucleotides may then combine with other nucleotides when one of the OH groups from the deoxyribose molecule and an OH group from a phosphate group from another molecule react, forming a new bond and water (Figure 25.18).

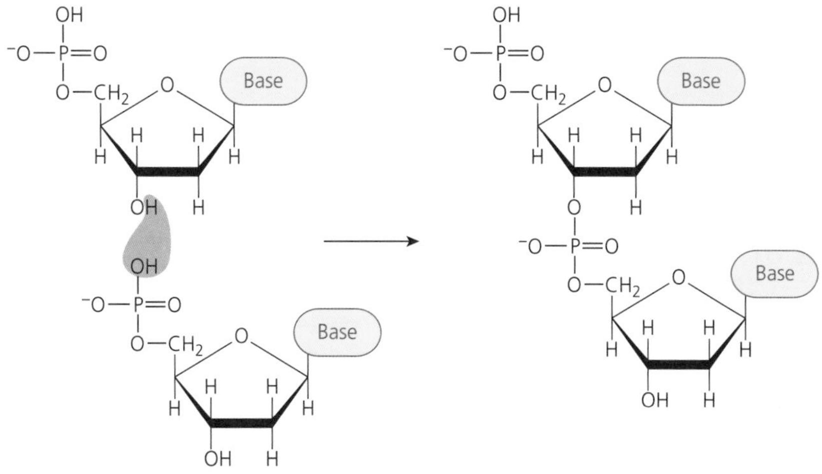

Figure 25.18 Bonding between nucleotides

This process repeats itself to extend the chain length so that eventually millions of nucleotides may bond together.

When a chain wraps around itself, bases can then hydrogen bond together to maintain the helical shape of the DNA molecule.

The bases

REVISED

A chemical base is a proton acceptor. So there must be lone pairs available on a donor atom that can then accept a proton by using a coordinate bond. The bases used by DNA are amines and have lone pairs available on nitrogen atoms that then hydrogen bond to another base.

There are four bases used to pair up in DNA:

- adenine and thymine (an A–T pair)
- guanine and cytosine (a G–C pair)

The reason why each base can pair up with its opposite base and no other relates to the number of hydrogen bonds that are possible between the bases, and the shapes of the bases involved.

In Figure 25.19(a) only two hydrogen bonds are possible as the best 'fit', whereas in Figure 25.19(b) three are possible.

(a) The A–T pair

Adenine
Thymine
Hydrogen bonds
Backbone of first chain
Backbone of first chain

(b) The G–C pair

Guanine
Cytosine
Backbone of first chain
Backbone of second chain
Note three hydrogen bonds this time

Figure 25.19 Base pairing

The action of cisplatin as an anti-cancer drug

REVISED

Cisplatin is a platinum(II) compound formed when two chloride ligands and two ammonia ligands bond to a central platinum(II) ion to form a square planar complex (Figure 25.20).

Figure 25.20 Cisplatin

This complex shows *E–Z* (cis-trans) isomerism as the two chloride ions (or ammonia molecules) occupy the same side of one of the edges of the complex. This complex has anti-cancer properties because it aims to prevent further replication taking place within the DNA strand that results in different forms of DNA, and these could result in cancerous growths. It is used to treat various types of cancers including sarcomas, some carcinomas, ovarian cancer, lymphomas, bladder cancer and germ cell tumours.

The mechanism by which it operates involves an initial attack by a water molecule to displace a chloride ion from cisplatin:

$$[Pt(NH_3)_2Cl_2] + H_2O \rightarrow [Pt(NH_3)_2Cl(H_2O)]^+ + Cl^-$$

The resulting $[Pt(NH_3)_2Cl(H_2O)]^+$ complex is then attacked by one of the nitrogen atoms from a guanine base from the DNA strand, removing the newly bound water molecule in another ligand displacement reaction:

$$Guanine_{(DNA)} + [Pt(NH_3)_2Cl(H_2O)]^+ \rightarrow Guanine_{(DNA)}\text{–}[Pt(NH_3)_2Cl]^+ + H_2O$$

Then another guanine from another strand removes the remaining chloride ion. The effect of this is that cisplatin now binds both strands together and stops further cancerous division and replication of the DNA strand.

It should be noted that cisplatin has many well-known side-effects, and these need considering before administering the compound to a patient. Kidney damage is a major concern that can limit its dose. Potential nerve damage and hearing loss can also be an issue.

Exam practice

1 Consider the three compounds drawn below, all of which are Brönsted–Lowry bases.

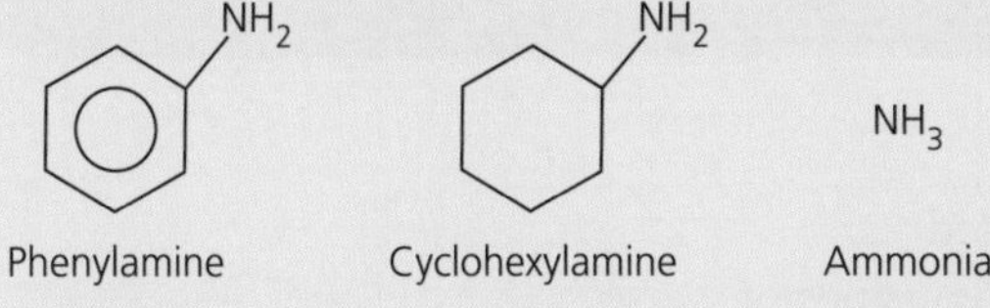

Phenylamine   Cyclohexylamine   Ammonia

- (a) State what is meant by the term 'Brönsted–Lowry base'. [1]
- (b) Write an equation to show cyclohexylamine acting as a base. [1]
- (c) Write the compounds in order of decreasing basic strength — strongest base first. [1]
- (d) Explain your answer to part (c). [3]
- (e) Draw the structure of the compound formed when phenylamine reacts with:
  - (i) bromomethane, CH_3Br [1]
  - (ii) ethanoic anhydride, $(CH_3CO)_2O$ [1]

2 The structure below shows a dipeptide formed by alanine and glycine. The shorthand form of this structural sequence is Ala–Gly; that is an alanine unit bonded to a glycine unit. It is shown in its zwitterionic form.

Ala-Gly

- (a) Draw the structure of the dipeptide called 'Gly–Ala'. [1]
- (b) Are the structures of Gly–Ala and Ala–Gly the same? Explain your answer. [2]
- (c) Draw the structures (showing all the charges on the appropriate ions) of the products formed when Ala–Gly is warmed with:
  - (i) NaOH(aq) [2]
  - (ii) HCl(aq) [2]

Answers and quick quiz 25 online

ONLINE

Summary

You should now have an understanding of:

- how amines can act as Brönsted–Lowry bases
- the structural features that determine base strength
- how amines react with halogenoalkanes
- how aliphatic amines can be prepared by reducing a nitrile
- how aromatic amines can be prepared by reduction of a nitro compound
- amino acids and their general structure
- zwitterions and how they are formed
- proteins and how they are formed from amino acids
- protein hydrolysis and how different products form depending on the pH at which the hydrolysis takes place
- the chemical structure of DNA in terms of its simple nucleotide building blocks
- the role of cis-platin as an anti-cancer compound, and how it works in terms of its interaction with the DNA strand

26 Polymers

Polymers are formed when many small molecules, units called **monomers**, bond together to form long chains. There are two types of polymer that can form depending on the original molecules — addition polymers and condensation polymers.

Addition polymers

Molecules containing carbon–carbon double bonds can be polymerised under certain conditions to form **addition polymers**. The π-bonding electrons from the double bond are used to join carbon atoms together, with the result that an extensive carbon chain is formed (Figure 26.1). This chain constitutes the backbone of the polymer.

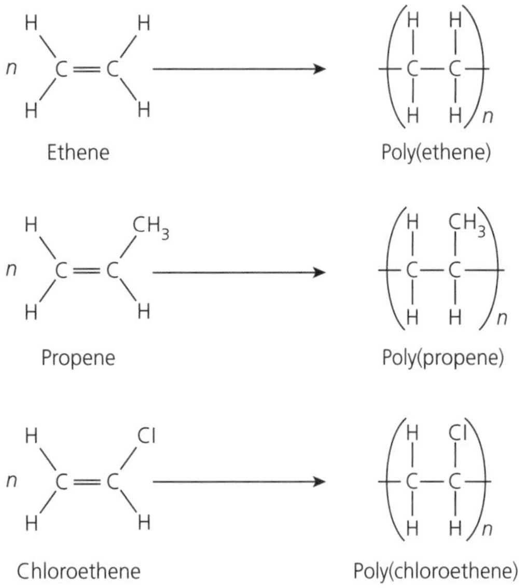

Ethene → Poly(ethene)

Propene → Poly(propene)

Chloroethene → Poly(chloroethene)

Figure 26.1 Addition polymerisation

If the structure of a polymer is given in a question, the monomer can be deduced by spotting the repeat unit (Figure 26.2), and then drawing out the monomer with its carbon–carbon double bond shown.

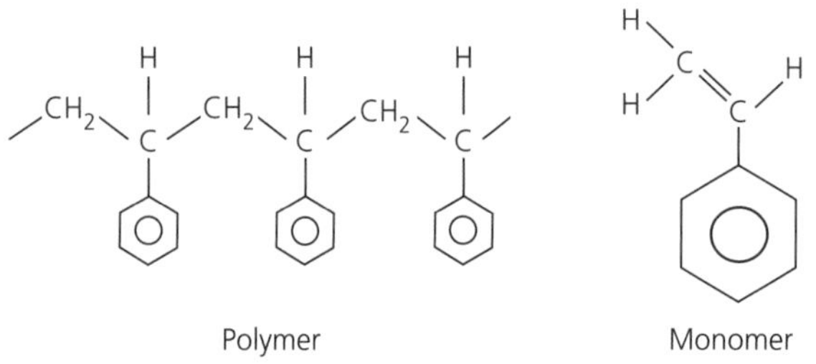

Polymer   Monomer

Figure 26.2 What is the monomer?

Now test yourself

TESTED

1 Draw the structures of the monomers that would polymerise to form the following addition polymers:

(a) $-CH(CONH_2)-CH_2-CH(CONH_2)-CH_2-CH(CONH_2)-CH_2-CH(CONH_2)-CH_2-CH(CONH_2)-CH_2-$

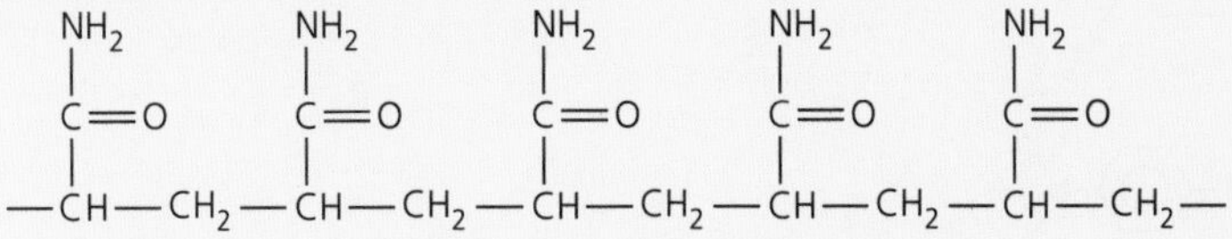

(b) $-CH(OCOCH_3)-CH_2-CH(OCOCH_3)-CH_2-CH(OCOCH_3)-CH_2-CH(OCOCH_3)-CH_2-CH(OCOCH_3)-CH_2-$

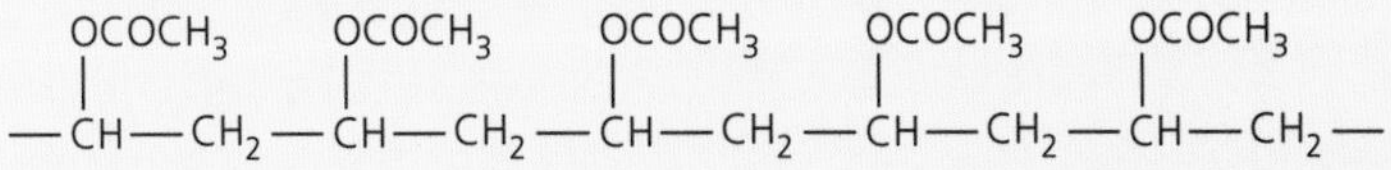

2 Draw the repeat unit of the polymer expected when this monomer polymerises:

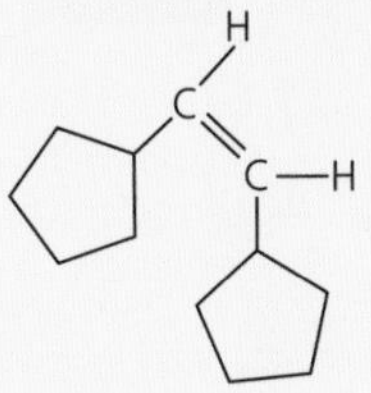

H(C₅H₉)C=C(C₅H₉)H

Answers on p. 223

Condensation polymers

Some molecules that have more than one functional group are able to link together and form long chains. A small molecule like water or hydrogen chloride is also formed in this type of polymerisation and the polymer, as a result, is called a **condensation polymer**.

Nylon (a polyamide) and **terylene** (a polyester) are examples of condensation polymers. When forming nylon (Figure 26.3) it is possible to use a dicarboxylic acid and a diamine.

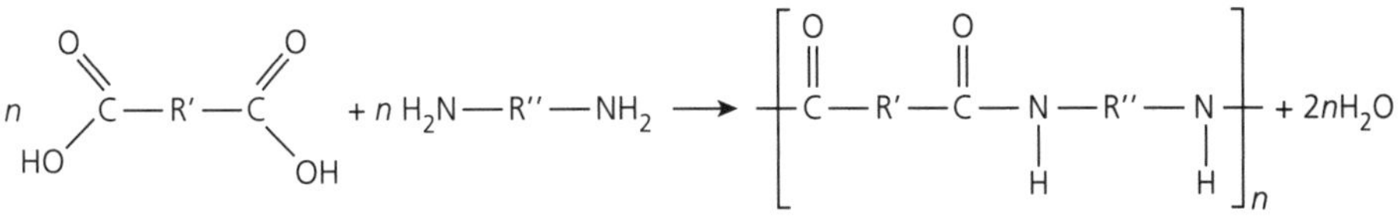

$$n\,HOOC-R'-COOH + n\,H_2N-R''-NH_2 \longrightarrow \left[-C(=O)-R'-C(=O)-N(H)-R''-N(H)-\right]_n + 2nH_2O$$

Figure 26.3 Condensation polymerisation making nylon

Terylene is a polyester and is made by the reaction of benzene-1,4-dicarboxylic acid with ethane-1,2-diol (Figure 26.4).

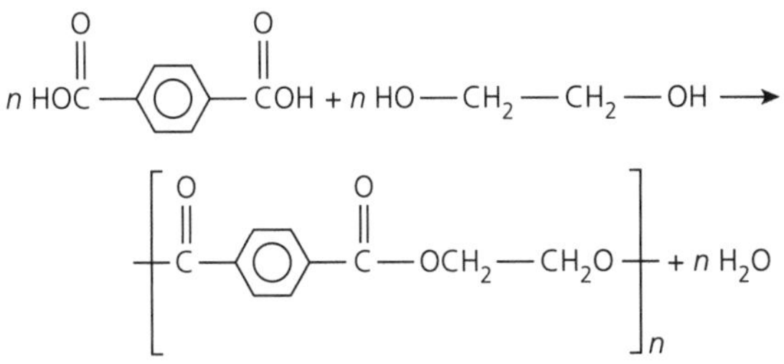

$$n\,HOC(=O)-C_6H_4-C(=O)OH + n\,HO-CH_2-CH_2-OH \longrightarrow \left[-C(=O)-C_6H_4-C(=O)-OCH_2-CH_2O-\right]_n + n\,H_2O$$

Figure 26.4 Making terylene using condensation polymerisation

Kevlar is a condensation polymer (a polyamide) that has been put to many interesting uses — for example in making bulletproof vests, in which strength and resistance to rapid, extreme forces are properties that make it particularly useful. It is made by reacting two monomers — a diacycl chloride and a diamine.

The interactions between two strands of Kevlar are shown in Figure 26.5. Notice how **hydrogen bonds** bind different strands of polymer together.

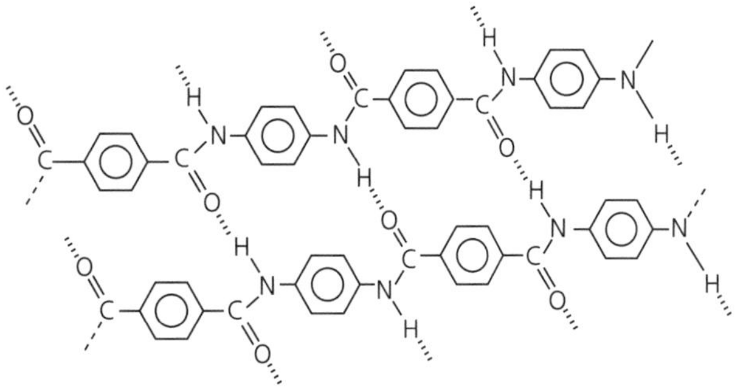

Figure 26.5 Kevlar

Now test yourself

TESTED

3 This is the structure of a condensation polymer.

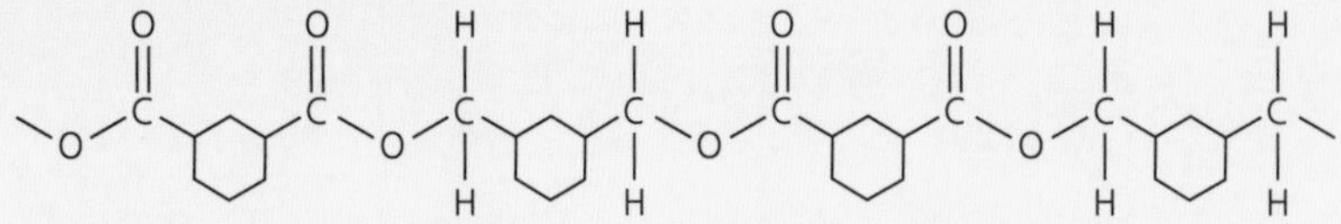

Draw the structures of the two monomers that would form this polymer.

4 The structure of the compound called phenylalanine is shown below.

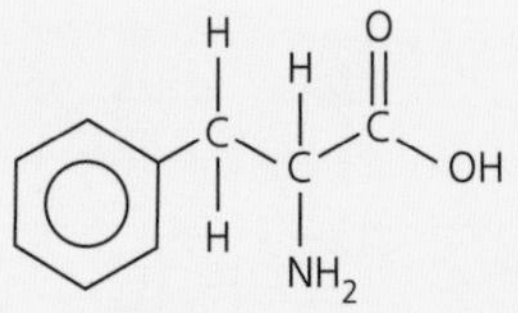

(a) What type of compound is phenylalanine?
(b) Draw the structure of the repeating unit that phenylalanine forms when polymerised.
(c) It is possible for phenylalanine to form a cyclic dimer in which two molecules combine, also producing two molecules of water. Draw the structure of this dimer.

Answers on p. 223

Disposing of polymers

When objects made of polymers come to the end of their useful lives, they are often disposed of in landfill waste sites. There are serious issues associated with this.

- Many addition polymers, like polyalkenes, are **chemically inert** — they contain strong carbon–carbon bonds and are non-polar. This means that possible nucleophiles, like water, will not decompose (**biodegrade**) these polymers even in acidic or basic conditions. As a result, many addition polymers have a long residence time during which they will only biodegrade over many centuries.
- Polar polymers, like polyesters and various nylons, are biodegradable. They are hydrolysed to form the constituent monomers, many of which are water soluble.
- Polymers can be burned instead of disposing of them in the ground. The **incineration** process involves high-temperature combustion and produces heat energy that can then be transformed into electrical energy (using steam turbines). Various pollutants also form in this process, many of which are potentially harmful to the environment. For example, toxic and highly acidic hydrogen chloride gas forms

if chlorinated polymers like polyvinylchloride are incinerated. Incineration also produces hazardous compounds called dioxins — for example, polychlorinated biphenyls (PCBs) — but modern incinerators can now reach very high temperatures, which destroy these dioxins.

- Polymers that melt on warming (thermoplastic polymers) can be remoulded and new products can be produced at a considerably lower energy cost than forming a polymer from crude oil via cracking etc. Sorting polymers prior to melting can be problematic because many different types are used.

Exam practice

1 Propene gas, under certain conditions, forms a waxy, white solid called poly(propene).
 (a) State the type of reaction that has taken place. [1]
 (b) Give the conditions required for this reaction to take place. [1]
 (c) Draw the displayed formula for propene. [1]
 (d) Draw a section of a poly(propene) molecule in which *three* propene molecules have joined. [2]

2 Kevlar is a polymer with the structure shown in Figure 26.5.
 (a) What type of polymer is Kevlar, addition or condensation? Explain your answer. [2]
 (b) Draw the *two* monomers that can react together to form Kevlar. [2]

3 Identify compounds V–Z, mentioned below, by drawing their displayed formulae (showing all bonds) and naming them.
 (a) Compound V has the molecular formula C_3H_6. It does not react with bromine water and all attempts to polymerise it fail. [1]
 (b) The polymer of compound W has this structure: [1]

 (c) Compounds X and Y polymerise to form this polymer unit: [2]

 (d) Compound Z is formed when the compound having molecules as shown below is heated with NaOH(aq) and then has HCl(aq) added. [1]

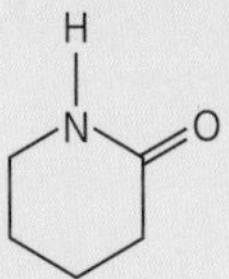

Answers and quick quiz 26 online

ONLINE

Summary

You should now have an understanding of:

- what addition polymers are
- how to draw the structure of an addition polymer given the monomer
- how to draw the structure of a monomer given the polymer
- what condensation polymers are
- how dicarboxylic acids react with diols to form polyesters like terylene
- how diamines react with dicarboxylic acids to form polyamides like nylon and Kevlar
- how polyesters and nylons can be hydrolysed to form their constituent monomers
- various disposal methods for polymers and how these depend on the chemical properties of polymers

27 Synthesis, analysis and structure determination

There are many modern analytical techniques that can be used to determine the structures of molecules. These techniques include infrared spectroscopy, mass spectrometry, nuclear magnetic resonance spectroscopy and chromatography.

Infrared spectroscopy

Atoms in molecules are **vibrating** about a fixed position. The frequency of the vibration varies according to the nature of the atoms bonded together:

- The *greater the mass* of the atoms, the lower the frequency of the vibrations.
- The *stronger the bond*, the higher the frequency of the vibrations.

The frequency of such vibrations can be quoted either in Hz or as the reciprocal of the wavelength in centimetres (called the **wavenumber** or the number of waves in 1 cm). A particular bond will vibrate or resonate at a particular frequency. The vibration frequency of some covalent bonds is given in Table 27.1.

Table 27.1 Bond vibration frequencies

Bond	Wavenumber/cm^{-1}
N–H (amines)	3300–3500
O–H (alcohols)	3230–3550
C–H	2850–3300
O–H (acids)	2500–3000
C≡N	2220–2260
C=O	1680–1750
C=C	1620–1680
C–O	1000–1300
C–C	750–1100

Figure 27.1 shows a typical infrared spectrum.

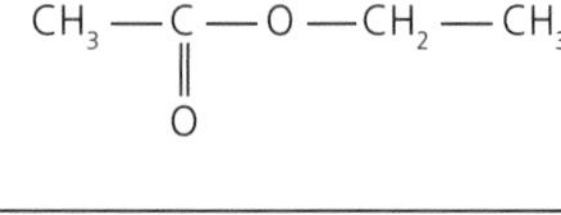

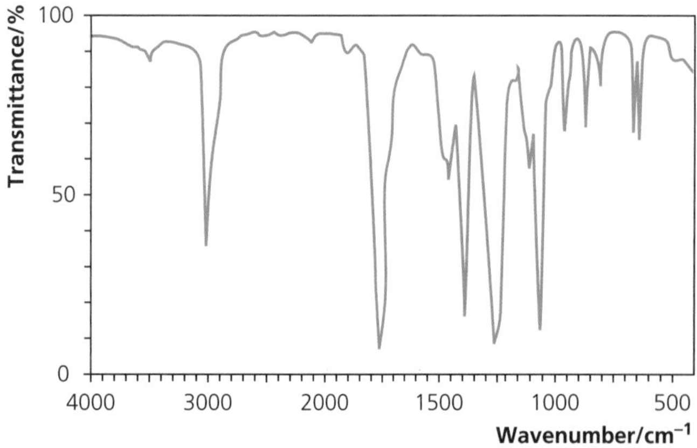

Figure 27.1 IR spectrum of ethyl ethanoate

Note the absorptions taking place at 1740 cm^{-1} (the C=O stretch in the ester) and also at 1240 cm^{-1} (the C–O stretch).

If we didn't know the identity of the substance, we could deduce that a C=O and a C–O bond were present, but this is all. However, the fingerprint region (the collection of absorptions below 1500 cm^{-1}) can often be used to identify a compound. This region of the spectrum can be compared to a database of spectra because the fingerprint region is unique for a particular molecule.

Mass spectrometry

A sample is injected into a mass spectrometer and a spectrum is produced that measures the **mass-to-charge ratio** (m/z) of a molecule (and its fragmented ions) and plots this against the corresponding **abundance** (the number of ions of a particular mass-to-charge ratio).

A molecule can undergo the following changes in the machine:

- Simple ionisation, in which a high-energy electron is removed from the molecule, M:

  $M(g) \rightarrow M^{+}(g) + e^{-}$

  M^+ is called the **molecular ion**, and an accurate measure of its relative molecular mass can be used to identify the molecule. The most useful peak in a mass spectrum is that of the molecular ion — seen as the most significant peak on the right of the spectrum.
- The molecular ion may undergo **fragmentation**:

  $M^{+}(g) \rightarrow X^{+}(g) + Y{\bullet}(g)$

  The charged species formed, X^+, will be detected and its m/z ratio measured. Some ions are particularly stable and their abundances tend to be relatively high — these include carbocations (R^+) and acylium ions (RCO^+).

Figure 27.2 shows a typical mass spectrum for $H_3CCOCH_2OCH_3$. Notice the following about this spectrum:

- The relative molecular mass of the compound is 88 because the peak furthest to the right is at m/z = 88.
- The peak at m/z = 43 is probably due to the fragment ion $[CH_3CO]^+$.
- The peak at m/z = 45 is probably due to the fragment ion $[CH_3OCH_2]^+$.

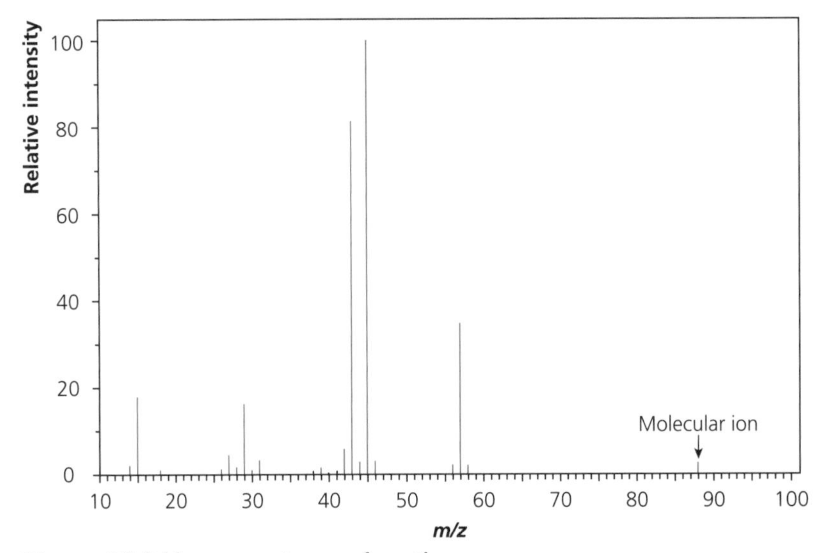

Figure 27.2 Mass spectrum of methoxypropanone

Now test yourself

TESTED

1 The infrared spectrum and mass spectrum for a compound, X, are given below.

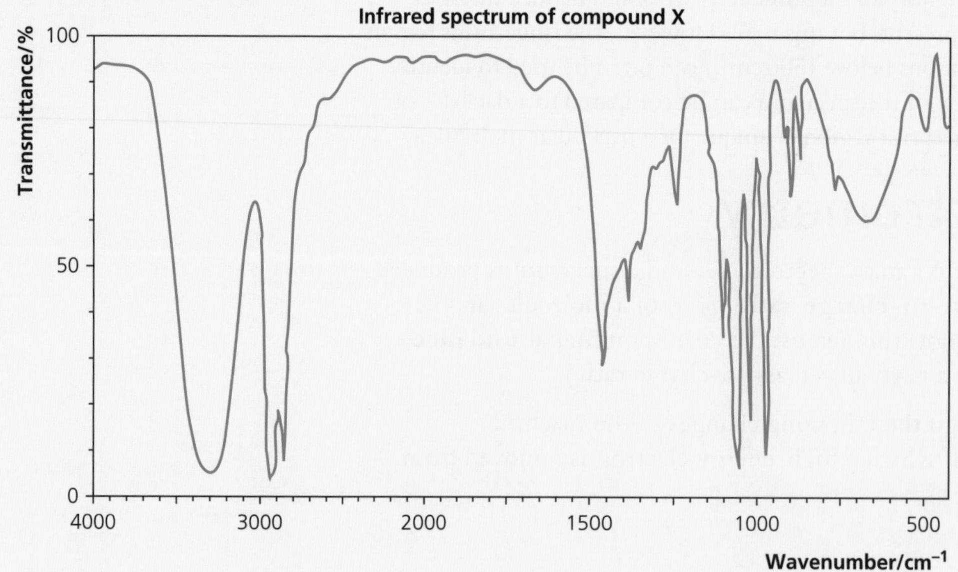

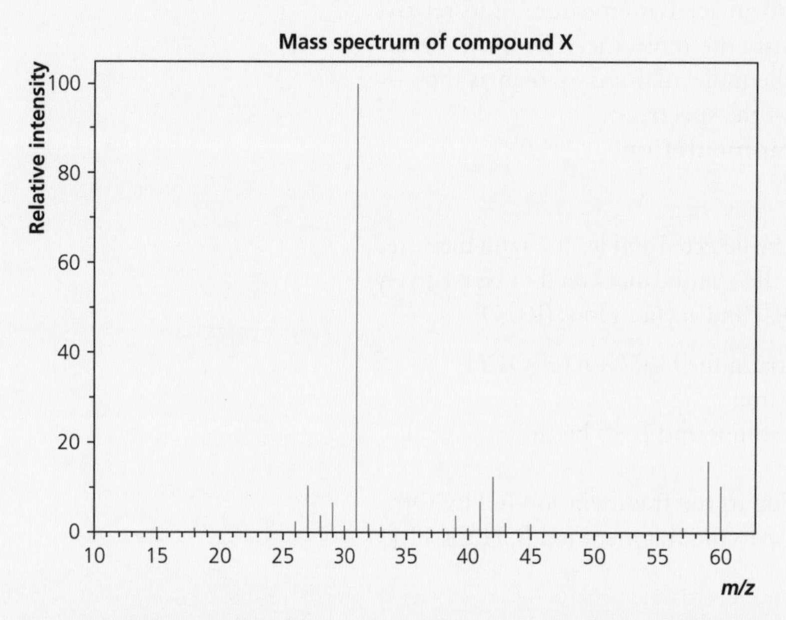

(a) Use the infrared spectrum and Table 27.1 to deduce which important bonds are present in X.
(b) Use the mass spectrum to determine the relative molecular mass of X.
(c) Given that X contains 60.0% carbon, 13.3% hydrogen and the rest is oxygen, calculate:
  (i) the empirical formula of X
  (ii) the molecular formula of X
(d) Suggest a possible structure for X.

Answer on pp. 223–224

Nuclear magnetic resonance spectroscopy

The magnetic properties of some atomic nuclei can be used to determine their environment in molecules that contain them. Absorptions can be detected across the radio wave range of electromagnetic radiation. These can be used to generate a nuclear magnetic resonance (n.m.r.) spectrum.

Exam tip

It is not necessary to understand the science behind the resonance processes in n.m.r. — it can be highly complex. But you must be able to appreciate the importance of various aspects to interpret an n.m.r. spectrum.

Nuclei with an odd number of nucleons (protons + neutrons) have a quality known as **nuclear spin**. Examples of nuclei that have this property include ^{1}H, ^{13}C and ^{19}F.

How to understand an n.m.r. spectrum

REVISED

The following aspects are important to understand when interpreting an n.m.r. spectrum:

- A compound called tetramethylsilane (TMS), $Si(CH_3)_4$, (Figure 27.3) is used as a reference to which other proton resonances are compared. It is used to calibrate the signals produced when analysing the compound under test.

  Figure 27.3 Tetramethylsilane, TMS

  TMS has four methyl groups — all its protons are the same, or **equivalent**, and therefore resonate at the same frequency in the magnetic field.

  A few drops of TMS are all that is required because it produces a strong signal, due to its 12 protons all resonating at the same frequency. TMS can be easily removed from the sample under test because it is volatile and chemically inert.
- The **chemical shift** (δ) axis on the spectrum produced measures the resonances at which protons in the sample occur in relation to the resonance of the protons in the TMS reference. It is measured in parts per million (ppm).
- The further to the left-hand side a peak occurs along the chemical shift axis (with higher δ values), the closer the protons (or ^{13}C atoms) are to electronegative groups, like oxygen, in the molecule.
- The chemical shift tells us about the **chemical environment** of the protons (or ^{13}C atoms) in the molecule.
- The vertical axis measures the absorptions of energy that are happening.
- The **areas under the peaks** in a proton n.m.r. spectrum give us information about the number of hydrogen atoms giving rise to the absorptions.

There are two types of n.m.r. to cover — ^{13}C n.m.r. and ^{1}H (or proton) n.m.r.

^{13}C spectra

REVISED

The chemical shifts for ^{13}C atoms, together with the corresponding groups giving rise to the absorptions, are shown in Figure 27.4.

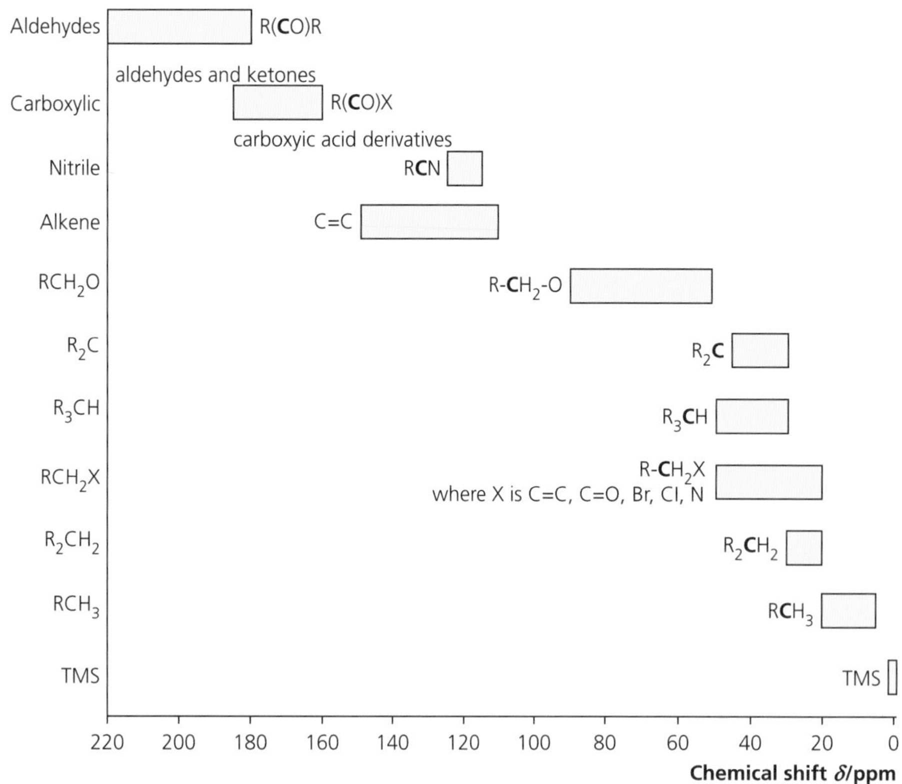

Figure 27.4 Chemical shifts

Figure 27.5 shows the ^{13}C spectrum for ethyl ethanoate.

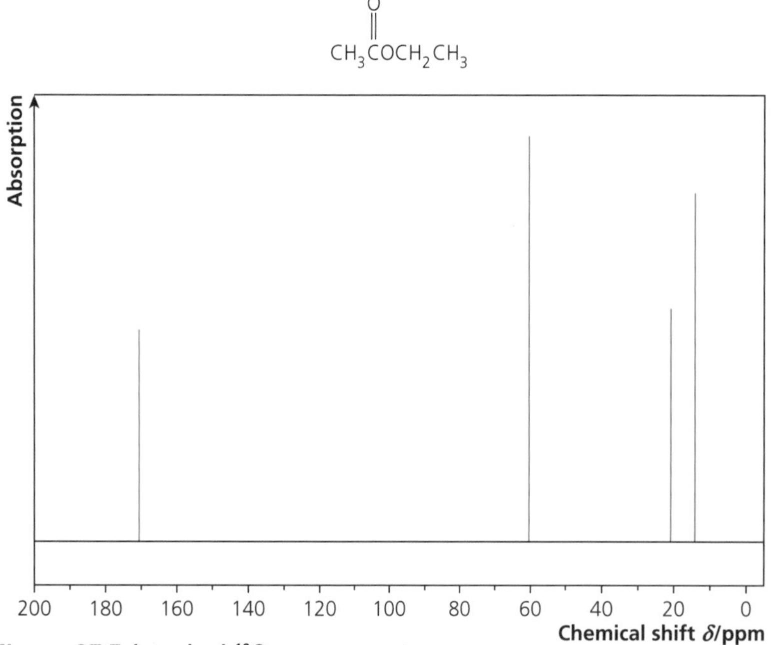

Figure 27.5 A typical ^{13}C n.m.r. spectrum

Note the following from Figure 27.5:

- The number of peaks tells us the number of different carbon atom environments in the molecule under test — four in this case.
- The peak at $\delta = 170$ is due to a carbonyl carbon because it is attached directly to an electronegative oxygen atom.
- The peak at $\delta = 60$ is due to a CH_2 carbon attached to the acyl group oxygen of the ester.
- The peaks at $\delta = 20$ and 15 are due to two CH_3 groups.

Exam tip

^{13}C n.m.r. spectroscopy is a lot more straightforward to understand than ^{1}H spectroscopy. It is often just a matter of counting the number of peaks (this gives the number of different types of carbon atom) and reading the positions of each peak along the chemical shift axis (this gives the chemical environment).

^{1}H spectra

REVISED

An ^{1}H n.m.r. spectrum is similar to a ^{13}C n.m.r. spectrum but the areas under the peaks and **spin–spin coupling** are extra considerations that make it a very powerful technique for deducing the identity of a molecule.

Spin–spin coupling happens when the spin of one proton couples with the spins of neighbouring *non-equivalent* protons and causes a signal to split.

Table 27.2 lists chemical shifts that will be of use in the questions that follow this section.

Table 27.2 Chemical shifts in ^{1}H n.m.r.

Type of proton	δ/ppm
ROH	0.5–5.0
RCH_3	0.7–1.2
RNH_2	1.0–4.5
R_2CH_2	1.2–1.4
R_3CH	1.4–1.6
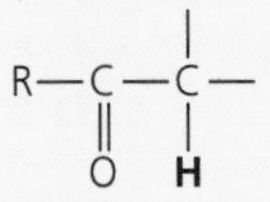 R—C(=O)—C(**H**)—	2.1–2.6
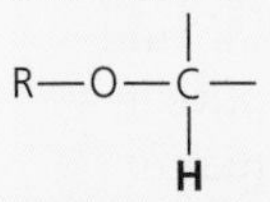 R—O—C(**H**)—	3.1–3.9
RCH_2cl or br	3.1–4.2
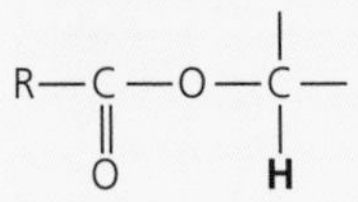 R—C(=O)—O—C(**H**)—	3.7–4.1
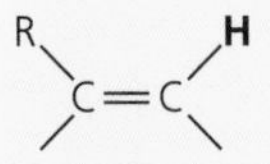 R(C=C)**H** (alkene C=C–**H**)	4.5–6.0
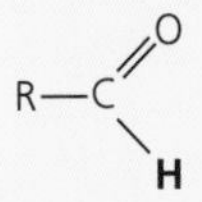 R—C(=O)—**H**	9.0–10.0
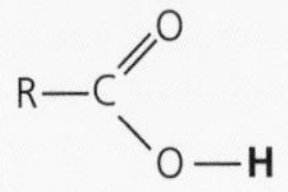 R—C(=O)—O—**H**	10.0–12.0

Figure 27.6 shows the ¹H n.m.r. spectrum for ethanol, C_2H_5OH.

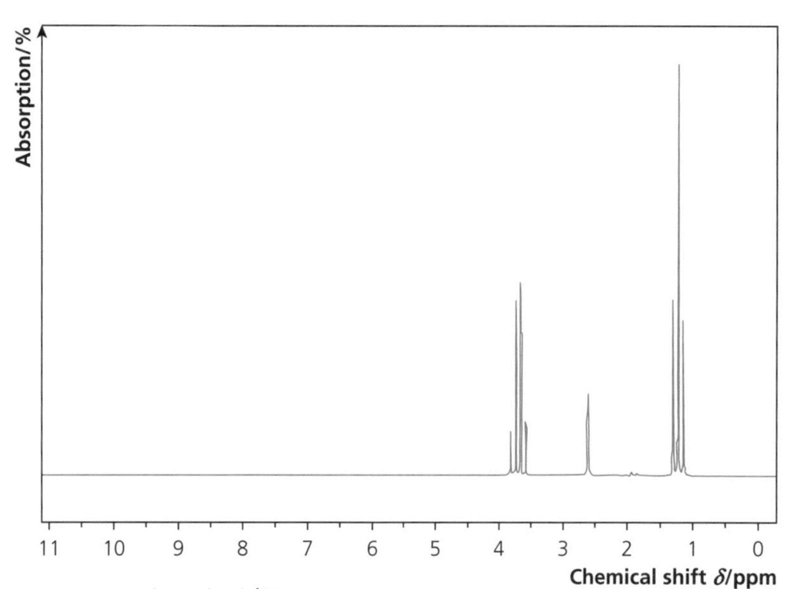

Figure 27.6 **A typical ¹H n.m.r. spectrum**

Important features of this spectrum include:

- three main peaks, so there are three proton environments
- a triplet at $\delta = 1.2$ due to the CH_3 protons
- a singlet at $\delta = 2.6$ due to the O–H proton
- a quartet at $\delta = 3.7$ due to the CH_2 protons

The areas under the peaks in Figure 27.6, reading from left to right, will have a ratio 2 : 1 : 3 because these are in proportion to the number of hydrogen atoms in each proton environment giving rise to each absorption.

Figure 27.6 also shows the displayed formula of ethanol.

- The CH_3 protons are next to the two CH_2 protons, and the CH_3 resonance is split by these **neighbouring protons**. The amount of splitting depends on the number of hydrogens 'next door' + 1, or $n + 1$. So, because there are two protons in the carbon atom next to the CH_3 group, the CH_3 resonance is split into 2 + 1 = 3, or a triplet.
- The pattern of these three split peaks can be judged from Pascal's triangle:

1

1 1

1 2 1

1 3 3 1

1 4 6 4 1

1 5 10 10 5 1

  Three numbers in a horizontal row correspond to 1 : 2 : 1. So a **1 : 2 : 1 triplet** is observed.
- The CH_2 protons are next to CH_3 protons, and so the CH_2 resonance will be split into 3 + 1 = 4 — a quartet is observed. The pattern of this quartet will be a **1 : 3 : 3 : 1 quartet**.
- The OH proton absorption is not split and this is the case in general — a proton attached to an electronegative atom like oxygen, does not experience any splitting from neighbouring protons. It therefore appears as a **singlet**.

Exam tip

Remember the 'n + 1' rule — the number of splits in an absorption is equal to the number of neighbouring protons + 1.

To conclude, the proton environments indicated in Figure 27.7 have now been associated with the adsorptions shown.

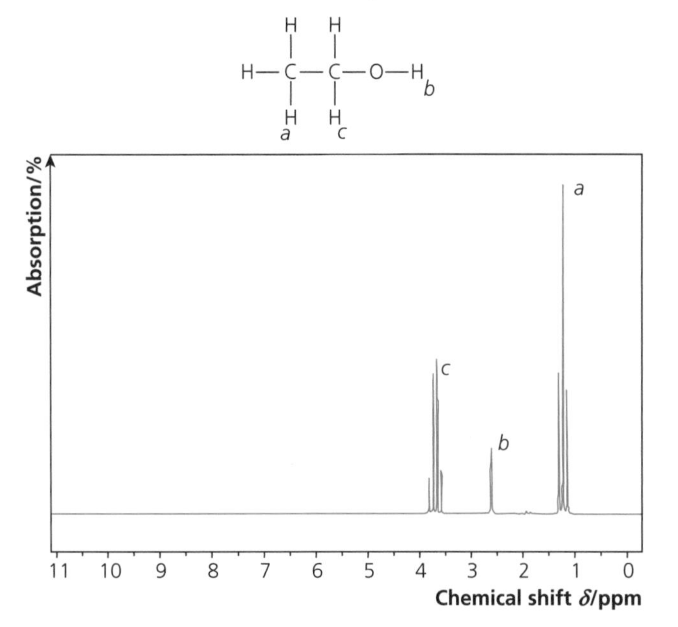

Figure 27.7 ^{1}H n.m.r. spectrum of ethanol

> **Exam tip**
>
> When considering ^{1}H and ^{13}C n.m.r. spectra in examination questions, remember to mention the number of peaks and the chemical shifts. In ^{1}H spectra you also need to consider areas under peaks and splitting patterns.

Now test yourself

TESTED

2 Consider the molecule represented by this displayed formula:

Predict and explain what its proton n.m.r. spectrum would look like.

Answer on p. 224

Chromatography

Gas–liquid chromatography can be used to separate mixtures of volatile liquids. Figure 27.8 shows a schematic diagram of the apparatus used.

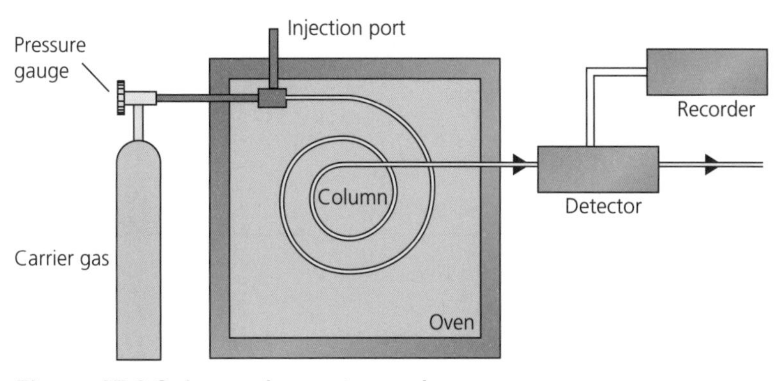

Figure 27.8 Column chromatography

The sample under test is injected into the column and heated. The oven consists of a long, narrow copper tube in which a material — for example, aluminium oxide — acts as the **stationary phase**. Many substances can be used as the stationary phase — some are polar, like aluminium oxide, and others are non-polar, like alkane-based materials. The sample mixture is moved through the stationary phase in the copper tube using an inert gas under pressure (**mobile phase**). The sample is separated into its individual components according to their **solubility** in the mobile phase and their **retention** in the stationary phase.

Types of chromatography include:

- thin-layer chromatography (TLC), in which a plate is coated with a solid and a solvent moves up the plate
- column chromatography (CC), in which a column is packed with a solid and a solvent moves down the column
- gas chromatography (GC), in which a column is packed with a solid or with a solid coated by a liquid, and a gas is passed through the column under pressure at high temperature

One substance that can be analysed using gas–liquid chromatography is premium-grade petrol. Figure 27.9 shows a typical chromatogram.

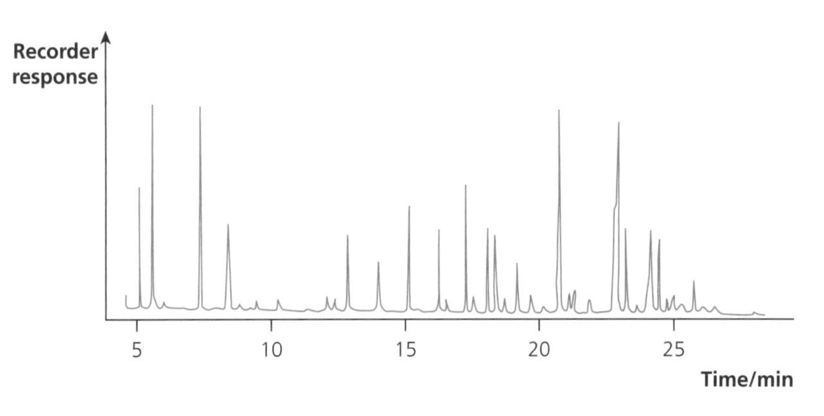

Figure 27.9 Typical gas–liquid chromatogram

Notice that there are many peaks, each representing a compound that comes out of the chromatography column. Even after a short retention time of a few minutes, compounds start to emerge from the column — these are relatively insoluble in the stationary phase but soluble in the mobile phase. Those that are retained in the column for 20–25 minutes are considerably more soluble in the stationary phase and take more time to be removed.

Separation depends on the balance between solubility in the moving phase and retention by the stationary phase.

A powerful method of analysis is to connect a chromatography technique, such as gas chromatography, to a mass spectrometer. The resulting separated compounds from the GC machine are then analysed individually by the mass spectrometer. This is known as GC–MS.

Exam tip

The length of time a compound spends in the column depends on its solubility in the stationary phase, and also how it is affected by the mobile phase. A polar species will spend a long time in the column if a polar stationary phase is used.

R_f value

REVISED

The R_f value is used to determine the identity of a compound. It is simply the distance moved by the compound divided by the distance moved by the solvent, and its value is constant for a particular substance.

In the chromatogram in Figure 27.10, the R_f values for substances A, B and C are calculated by measuring the distance each compound moves up the stationary phase divided by the distance moved by the solvent (4.7 cm). M is a mixture containing three compounds. The R_f values of these compounds can then be compared with known R_f values to determine their chemical identity.

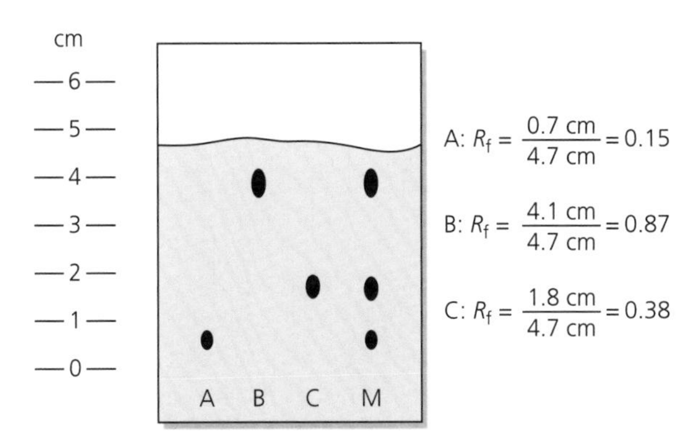

Figure 27.10 Determining R_f values

Exam practice

1 An ester, Y, is analysed to produce the three spectra shown below.

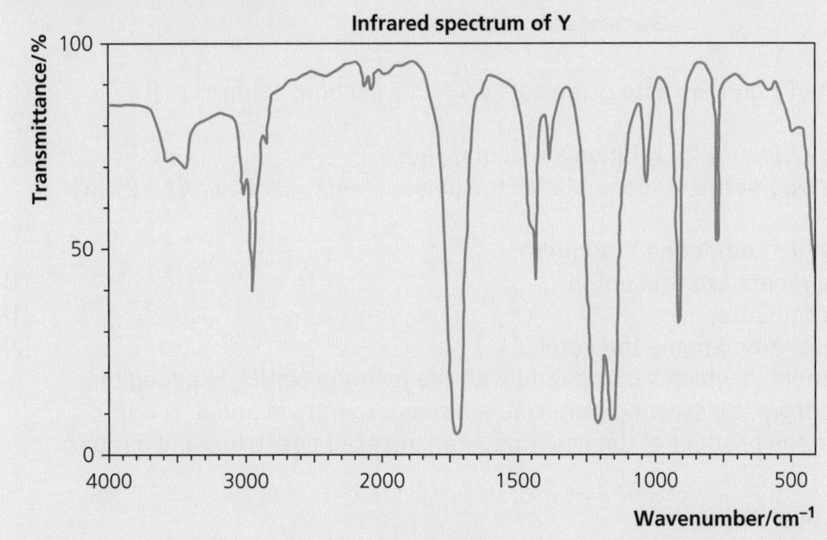

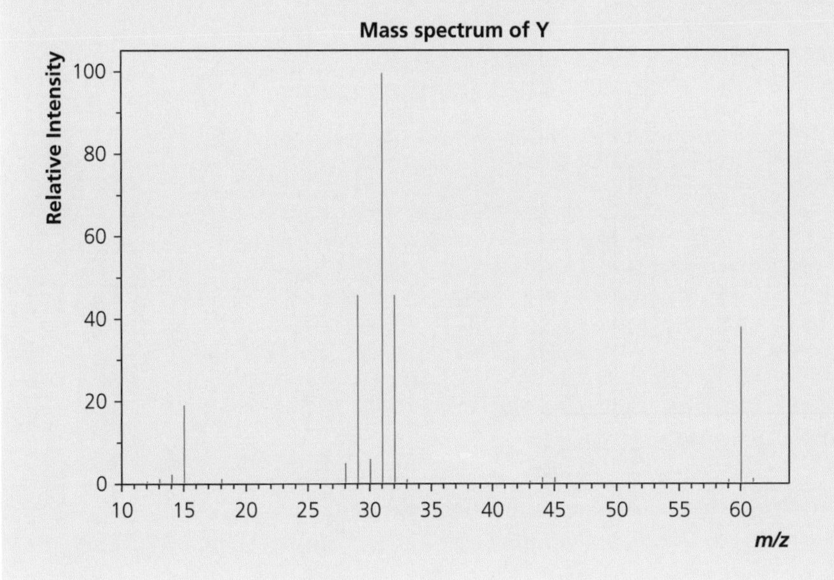

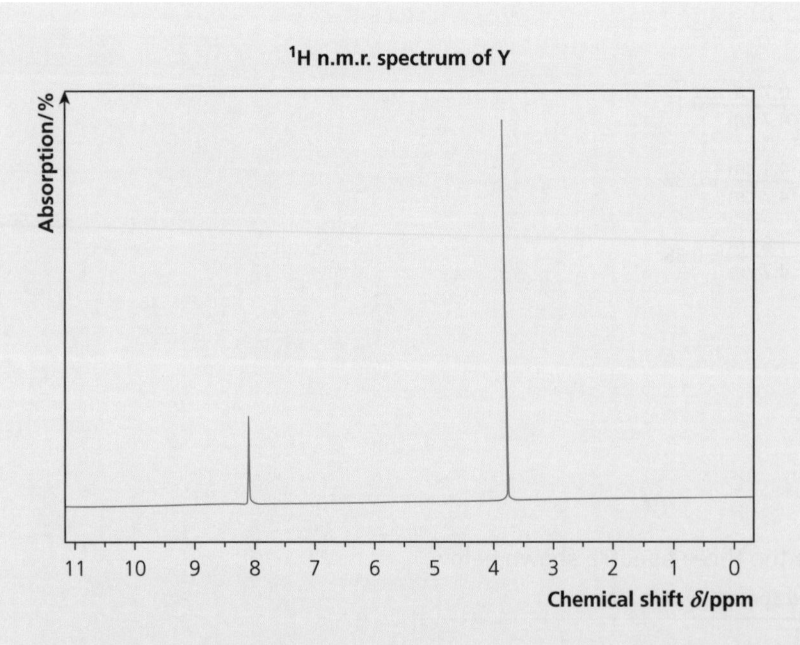

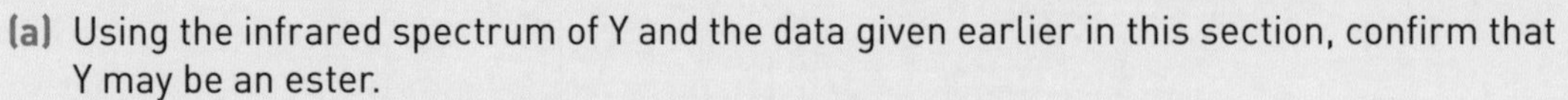

(a) Using the infrared spectrum of Y and the data given earlier in this section, confirm that Y may be an ester. [2]

(b) Using the mass spectrum of Y, deduce its relative molecular mass. [1]

(c) Suggest the formulae of the fragments responsible for the peaks at $m/z = 31$ and $m/z = 29$ in the mass spectrum of Y. [2]

(d) Using the ^{1}H n.m.r. spectrum for compound Y, suggest:

(i) how many proton environments are present in Y [1]

(ii) the nature of these environments [1]

(e) Draw the displayed formula for Y and name the ester. [2]

2 A student carries out an experiment in which a compound with the formula $NaBH_4$ is added to ethanal. He obtains two spectra from the starting material — a mass spectrum and a ^{1}H n.m.r. spectrum; and two spectra from the product of the reaction — an infrared spectrum and a mass spectrum.

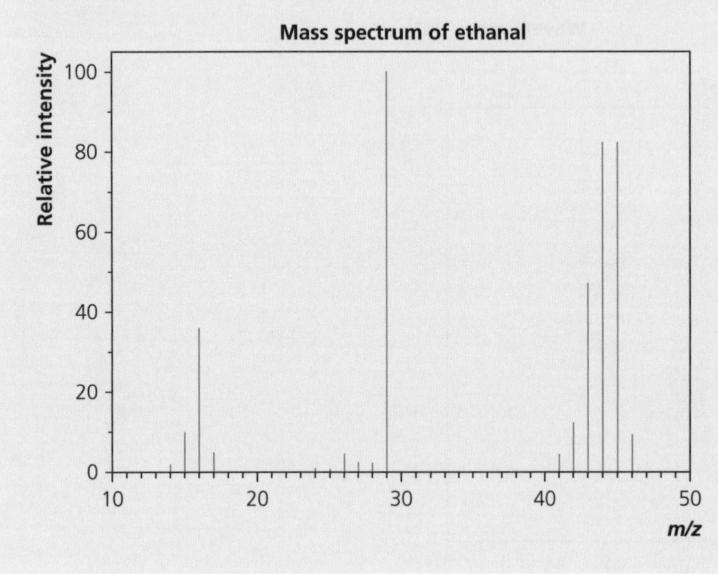

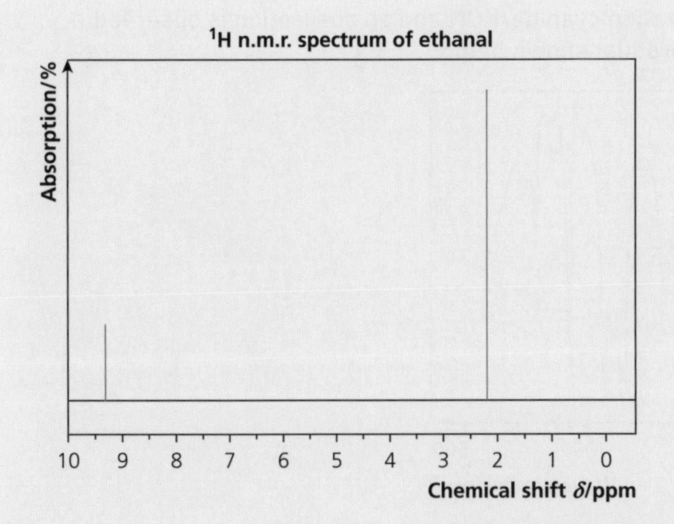

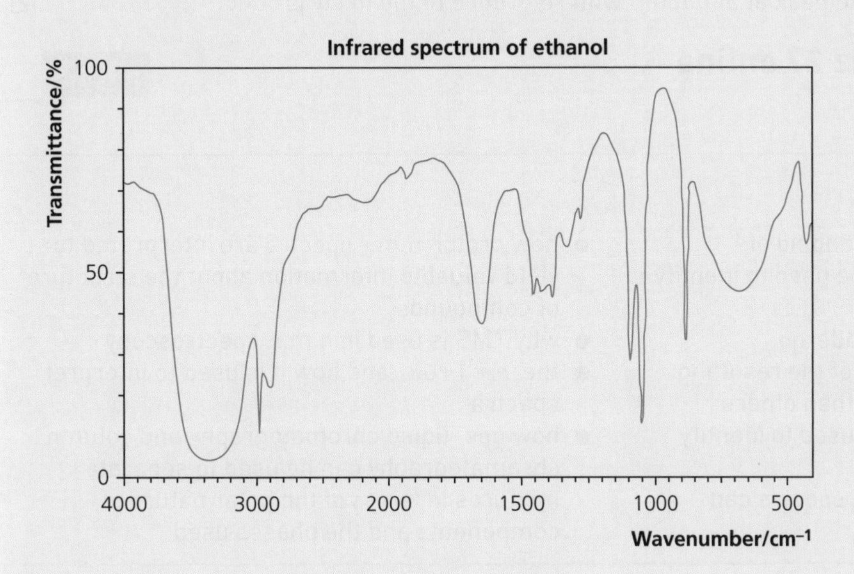

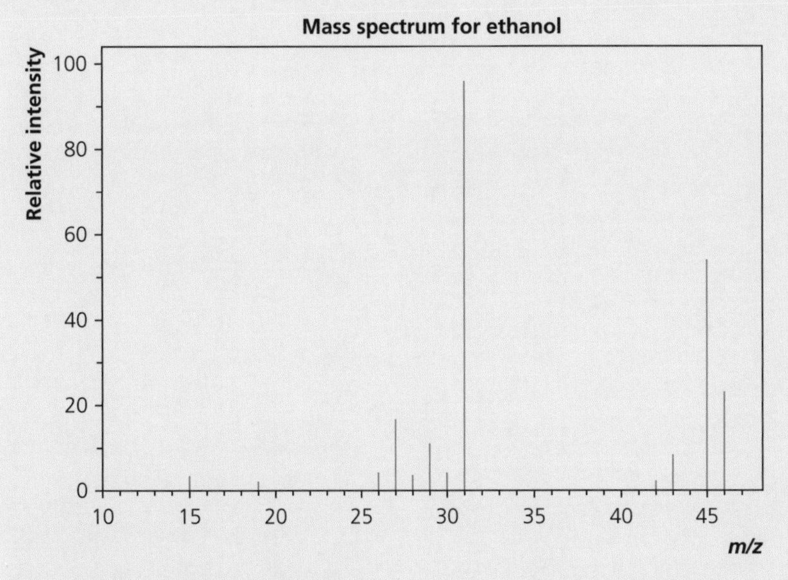

(a) Use the mass spectrum for ethanal to deduce its relative molecular mass. Explain how you arrived at your answer. [2]

(b) What is the species giving rise to the peak at $m/z = 29$? [1]

(c) Identify the proton environments giving rise to the adsorptions in ethanal's n.m.r. spectrum. [2]

(d) Explain how ethanol's infrared spectrum indicates that an alcohol has been produced. [1]

(e) A peak occurs at $m/z = 31$ in ethanol's mass spectrum. Explain the existence of this peak. [1]

In a further reaction, ethanal reacts with hydrogen cyanide/KCN and an absorption is observed at 3405 cm^{-1} in the infrared spectrum of the product shown below.

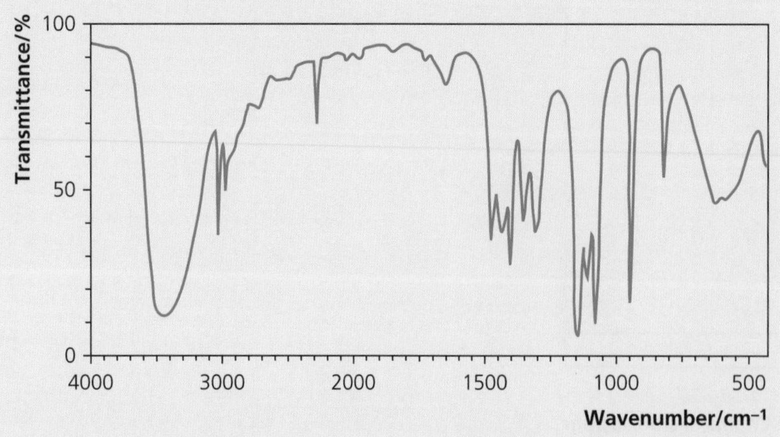

(f) Explain the existence of the peak at 3405 cm^{-1} with reference to the likely product. [2]

Answers and quick quiz 27 online

ONLINE

Summary

You should now have an understanding of:

- how mass spectroscopy can be used to identify compounds
- the fact that molecules can undergo fragmentation and that some of the resulting positive ions are more stable than others
- how infrared spectroscopy is used to identify specific bonds in molecules
- the information that ^{13}C n.m.r. spectra can provide
- how proton n.m.r spectra are interpreted to yield valuable information about the structure of compounds
- why TMS is used in n.m.r. spectroscopy
- the '*n* + 1 rule' and how it is used to interpret spectra
- how gas–liquid chromatography and column chromatography can be used to separate mixtures in terms of the polar nature of components and the phases used

Now test yourself answers

Chapter 1

1 (a) 9 protons, 10 neutrons, 9 electrons
(b) 34 protons, 40 neutrons, 34 electrons
(c) 22 protons, 26 neutrons, 20 electrons
(d) 35 protons, 44 neutrons, 36 electrons

2 relative atomic mass of krypton = $\left(\frac{0.35}{100} \times 78\right) + \left(\frac{2.3}{100} \times 80\right) + \left(\frac{11.6}{100} \times 82\right) + \left(\frac{11.5}{100} \times 83\right) + \left(\frac{56.9}{100} \times 84\right) + \left(\frac{17.4}{100} \times 86\right)$
= 83.93 (no units)

3 Let y = % of ^{10}B, then % of ^{11}B = 100 − y
Therefore, $\left(\frac{y}{100} \times 10\right) + \left(\frac{100 - y}{100} \times 11\right) = 10.8$
So, $0.1y + 11 - \left(\frac{11}{100}y\right) = 10.8$
This gives $0.2 = 0.01y$, so y = 20%, so ^{10}B = 20% and ^{11}B = 80%

4 (a) beryllium: $1s^2, 2s^2$; magnesium: $1s^2, 2s^2, 2p^6, 3s^2$; calcium: $1s^2, 2s^2, 2p^6, 3s^2, 3p^6, 4s^2$
(b) The number of electrons in the +3 ion is 10, so the atomic number is 10 + 3 = 13.

5 (a)
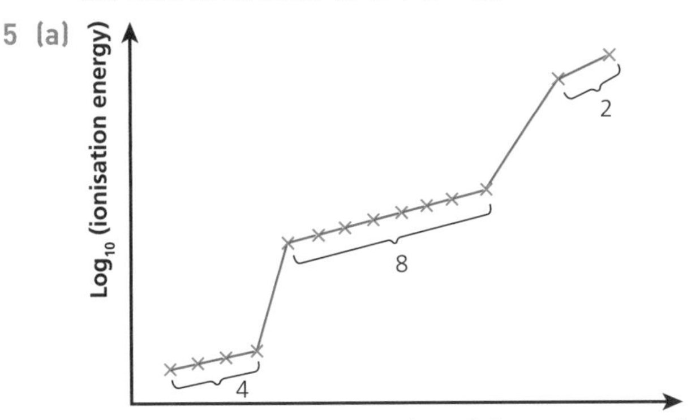

(b) $P^{3+}(g) \rightarrow P^{4+}(g) + e^-$

Chapter 2

1 amount of Ca = $\frac{1.00}{40.1}$ = 0.0249 mol
amount of Cl = $\frac{1.77}{35.5}$ = 0.0500 mol
ratio is 1 : 2; empirical formula is $CaCl_2$

2 amount of calcium = $\frac{0.210}{40.1} = 5.24 \times 10^{-3}$
amount of silicon = $\frac{0.147}{28.1} = 5.23 \times 10^{-3}$ mol
amount of oxygen = $\frac{0.252}{16.0}$ = 0.0158 mol
ratio is 1 : 1 : 3; empirical formula is $CaSiO_3$

3 (a) Assuming a total mass of 100 g:
amount of nitrogen = $\frac{30.4}{14.0}$ = 2.17 mol
amount of oxygen = $\frac{100 - 30.4}{16.0}$ = 4.35 mol
ratio is 1 : 2; empirical formula is NO_2
(b) Because the relative molecular mass is 92, twice the empirical formula is required; the molecular formula is N_2O_4.

4 (a) The molecular formula of caffeine is $C_8H_{10}N_4O_2$.
(b) (i) Relative molecular mass = (8 × 12.0) + (10 × 1.0) + (4 × 14.0) + (2 × 16.0) = 194
Given that the mass of caffeine is 5.60×10^{-3} g:
amount present in mol = $\frac{5.60 \times 10^{-3}}{194}$
= 2.89×10^{-5} mol
number of molecules this represents = $2.89 \times 10^{-5} \times 6.0 \times 10^{23} = 1.73 \times 10^{19}$
(ii) Each caffeine molecule consists of eight carbon atoms. So when 1.73×10^{19} caffeine molecules are completely combusted, the number of carbon dioxide molecules formed will be ($8 \times 1.73 \times 10^{19}$), or 1.39×10^{20} molecules.

5 Using $pV = nRT$:
$p \times 0.0330 = 0.905 \times 8.31 \times (200 + 273)$
pressure = 1.08×10^5 Pa

6 amount of Mg = $\frac{2.00}{24.3}$ = 0.0823 mol
amount of hydrogen = 0.0823 mol
volume of hydrogen = $\frac{nRT}{p} = \frac{0.0823 \times 8.31 \times 298}{100\,000}$
= 0.00204 m^3 or 2.04 dm^3

7 ratio of nitrogen to hydrogen is 1 : 3
volume of hydrogen required = 150 cm^3 × 3 = 450 cm^3

8 amount of H_2O_2 dissolved = $\frac{100}{1000} \times 0.500$ = 0.0500 mol
amount of oxygen formed = $\frac{0.0500}{2}$ = 0.0250 mol
volume = $\frac{nRT}{p} = \frac{0.0250 \times 8.31 \times 298}{100\,000}$ = 0.619 dm^3

9 Using moles dissolved = $\frac{\text{volume (cm}^3)}{1000\,\text{cm}^3}$ × concentration (in mol dm^3):

(a) amount dissolved = $\frac{10.0}{1000} \times 0.200\,\text{mol dm}^{-3}$
= 2.00×10^{-3} mol NaOH

(b) amount dissolved = $\frac{250}{1000} \times 1.20\,\text{mol dm}^{-3}$
= 0.300 mol HNO_3

10 amount of dissolved HCl = $\frac{20.0}{1000} \times 0.900$
= 0.0180 mol

amount of NaOH reacting is 0.0180 mol

So volume reacting = $0.018 \times \frac{1000}{0.0500} = 360\,\text{cm}^3$

11 atom economy = $\left(\frac{50.5}{50.5 + 36.5}\right) \times 100 = 58.0\%$

12 (a) NaF

(b) K_2SO_4

(c) $Al(OH)_3$

13 $2Al(s) + 3CuSO_4(aq) \rightarrow Al_2(SO_4)_3(aq) + 3Cu(s)$

14 $KOH(aq) + HCl(aq) \rightarrow KCl(aq) + H_2O(l)$

$H^+(aq) + OH^-(aq) \rightarrow H_2O(l)$

Chapter 3

1 (a)

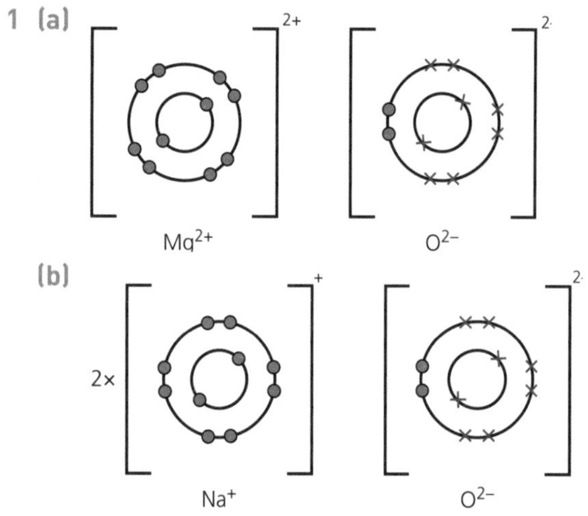

2 Lithium fluoride has a giant ionic lattice structure; there will be oppositely charged ions — Li^+ and F^-; these will be attracted to each other strongly by electrostatic forces; this means that they will be difficult to separate.

3 (a)

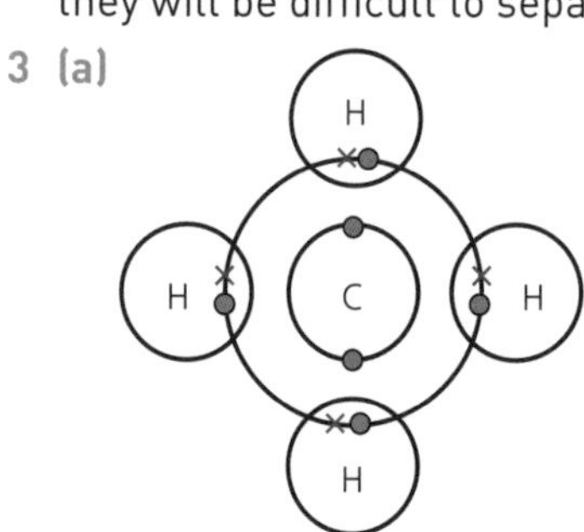

(b)

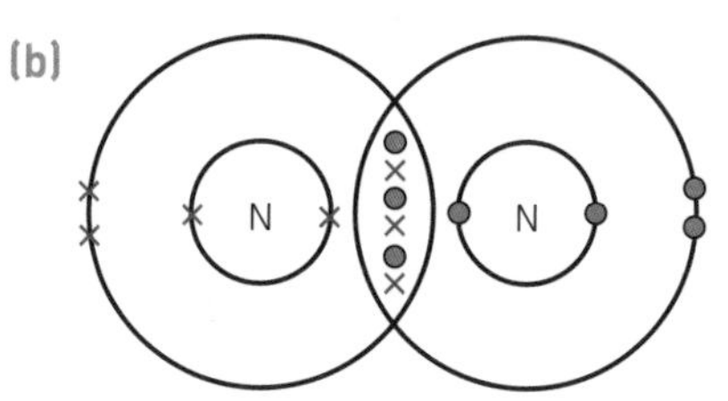

4 Methane is simple covalent; this means that the individual molecules will be relatively easy to separate as the substance melts.

5

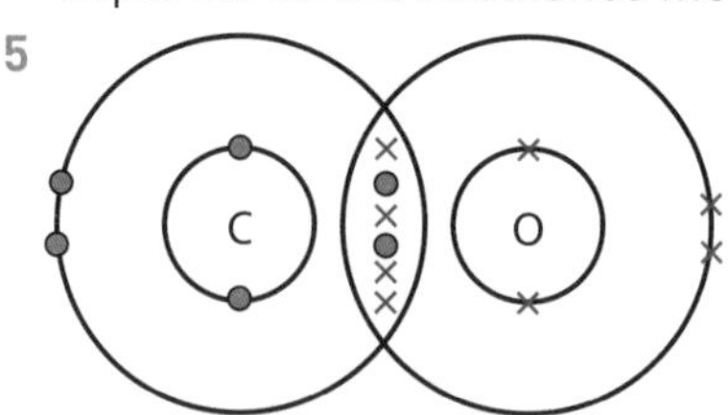

6 Hydrogen and fluorine have very different electronegativities; this results in a dipole forming in HF with the bonded electrons closer to the fluorine end of the molecule. In hydrogen, the atoms are identical; so their electronegativities will be the same; H_2 will therefore be a non-polar molecule.

7

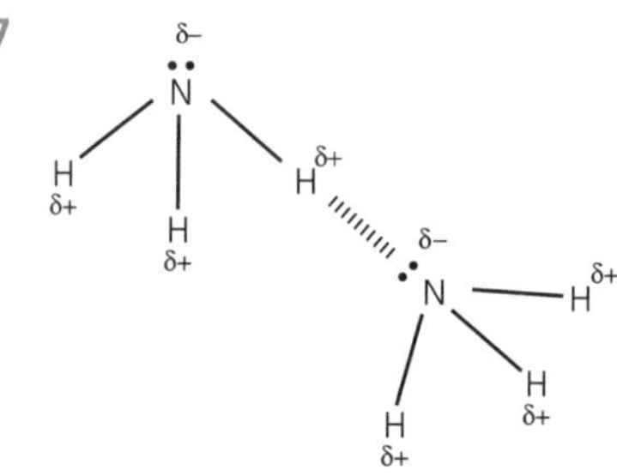

8 As the group is descended, molecules contain more electrons; therefore, there will be greater van der Waals forces; this means that it will be more difficult to separate the halogen molecules at the bottom of the group.

9 NaF has a giant ionic lattice structure whereas F_2 is simple covalent; in NaF the oppositely charged ions will be difficult to separate because they are electrostatically attracted; in F_2 the individual molecules are easy to separate; so fluorine has a lower melting point.

10 (a) tetrahedral

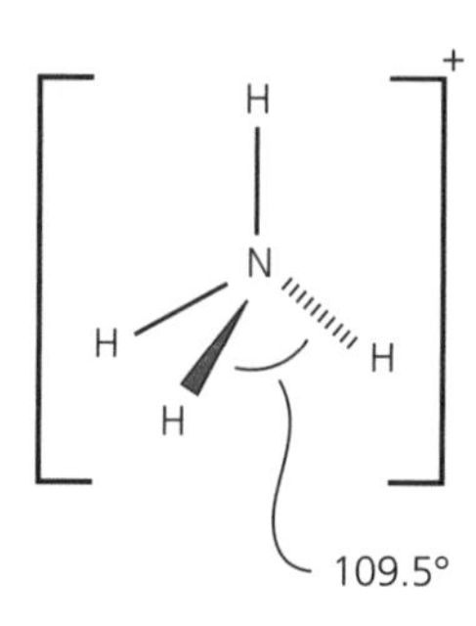

(b) trigonal planar

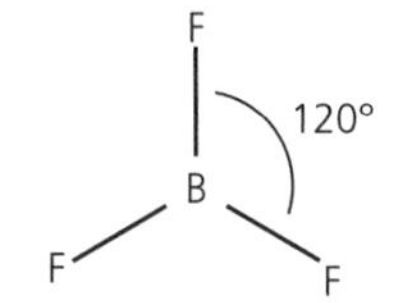

(c) octahedral

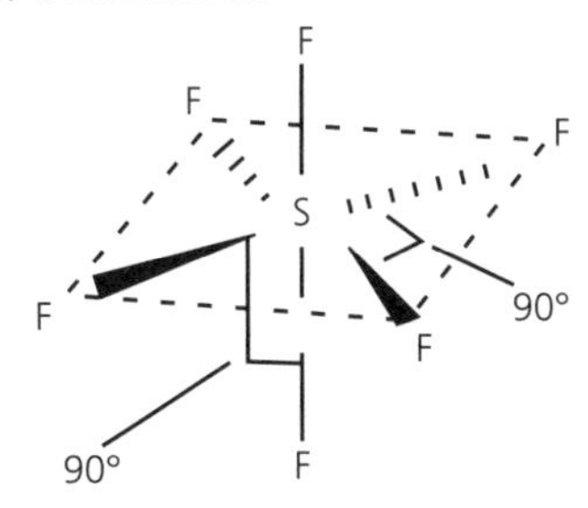

Chapter 4

1 Using $q = mc\Delta T$:

$q = 25.0 \times 4.18 \times 11.4 = 1191\,J = 1.19\,kJ$

amount of zinc used $= \frac{0.95}{65.4} = 0.0145\,mol$

enthalpy change $= \frac{1.19\,kJ}{0.0145\,mol} = -82.1\,kJ\,mol^{-1}$

2 $q = mc\Delta T$

$q = 100 \times 4.18 \times (25.0 - 17.0) = 3344\,J = 3.34\,kJ$

0.16 g of methanol (CH_3OH) is $\frac{0.16}{32.0}$ mol, i.e. 5.00×10^{-3} mol

enthalpy change $= \frac{3.34\,kJ}{5.00 \times 10^{-3}} = -668\,kJ\,mol^{-1}$

3 (a) $C(s) + O_2(g) \rightarrow CO_2(g)$

(b) $6C(s) + 7H_2(g) \rightarrow C_6H_{14}(l)$

4 $\Delta H = \Sigma\Delta_f H(\text{products}) - \Sigma\Delta_f H(\text{reactants}) = (4 \times -1279) - [-2984.0 + (6 \times -285.8)] = -5116 - (-4698.8) = -417.2\,kJ\,mol^{-1}$

5 $\Delta_f H(\text{propane}) = (3 \times -393.5) + (4 \times -285.8) - (-2220.0) = -1180.5 - 1143.2 + 2220.0 = -103.7\,kJ\,mol^{-1}$

6 ΔH = energy required to break bonds – energy released on forming new bonds = [409 + (2 × 388) + 436)] – [163 + (4 × 388)] = 1621 – 1715 = $-94\,kJ\,mol^{-1}$

Chapter 5

1 As temperature increases, particles move with a higher kinetic energy; and a greater proportion have an energy that exceeds the activation energy; so there will be more effective collisions per unit time; and the rate increases.

2 As the concentration increases, the number of ions per unit volume also increases; there will be more collisions per unit time; and the rate of reaction increases.

3 Manganese(IV) oxide is a catalyst for the decomposition of hydrogen peroxide to form water and oxygen. The catalyst provides an alternative reaction pathway, which is faster because it has a lower activation energy.

Chapter 6

1 (a) $K_c = \frac{[NH_3(g)]^2}{[N_2(g)]\,[H_2(g)]^3}$

(b) $K_c = \frac{[HI(g)]^2}{[H_2(g)]\,[I_2(g)]}$

(c) $K_c = \frac{[NO(g)]^2}{[N_2(g)]\,[O_2(g)]}$

2 (a) When the temperature is increased, the equilibrium position shifts to the left-hand side by using its endothermic route.

(b) As temperature increases, the rate of reaction will always increase; more molecules have an energy that exceeds the activation energy; there will be more effective collisions taking place per unit time; and the rate will increase.

(c) $K_c = \frac{[N_2O_4(g)]}{[NO_2(g)]^2}$

3

	Equilibrium position	Rate at which equilibrium is attained
Total pressure is increased	Shifts to the right-hand side	Increases
Temperature is increased	Shifts to the left-hand side	Increases
A catalyst is added	No change	Increases

Chapter 7

1 (a) +3

(b) +1

(c) +4

(d) +6

(e) +6

(f) +5

(g) +6

2 (a) cobalt(III) chloride

(b) sodium chlorate(I)

(c) titanium(IV) chloride

(d) sodium ferrate(VI)

(e) sulfuric(VI) acid

(f) iodate(V) ion

(g) manganate(VI) ion

3 oxidising agent: oxygen gas, O_2; reducing agent: SO_2

4 Pb: +4 to +2; Pb(IV) is the oxidising agent
Cl: −1 to 0; Cl^- is the reducing agent (some Cl^- is unchanged)

5 (a) $Zn(s) \rightarrow Zn^{2+}(aq) + 2e^-$
$Fe^{2+}(aq) + 2e^- \rightarrow Fe(s)$
(b) $2Al(s) \rightarrow 2Al^{3+}(aq) + 6e^-$
$3Cu^{2+}(aq) + 6e^- \rightarrow 3Cu(s)$

Chapter 8

1 p block

2 p block

3 (a) Magnesium atoms have 12 protons in the nucleus; sodium has 11; both have the same number of energy levels or shells. The outer electrons in magnesium will experience a stronger attraction; there will be a contraction in size, giving a smaller atomic radius.
(b) In aluminium, the electron being removed comes from a 3*p* orbital, which is slightly further away from the nucleus than the 3*s* orbital in magnesium from which its electron is removed; there is also slightly more shielding; so the electron being removed from an aluminium atom is attracted more weakly; the first ionisation energy will be lower.
(c) Both are simple covalent; but sulfur consists of S_8 molecules whereas chlorine consists of Cl_2 molecules; S_8 molecules have more electrons; and the van der Waals forces are greater; so its boiling point will be higher.

Chapter 9

1 $Ca(s) + 2H_2O(l) \rightarrow Ca(OH)_2(aq) + H_2(g)$
Bubbles would be observed; calcium metal (a light-grey solid) would eventually dissolve to form a white suspension in water.

2 Add barium chloride solution and hydrochloric acid to separate solutions of sodium sulfate and sodium nitrate; a white precipitate forms with sodium sulfate; there is no reaction with sodium nitrate.

3 Chlorine has fewer electron energy levels compared with iodine; so the added electron will be closer to the nucleus in the chlorine atom; and the electrostatic attraction will be greater; so chlorine will be the more powerful oxidising agent.

4 (a) $Br_2(aq) + 2NaI(aq) \rightarrow 2NaBr(aq) + I_2(aq)$ *or* $Br_2(aq) + 2I^-(aq) \rightarrow 2Br^-(aq) + I_2(aq)$
(b) Bromine has been reduced; iodide ions have been oxidised.

5 $KF(s) + H_2SO_4(l) \rightarrow KHSO_4(s) + HF(g)$

6 (a) H_2S: −2; SO_2: +4; S: 0
(b) $6HI(g) + H_2SO_4(l) \rightarrow 3I_2(s) + S(s) + 4H_2O(l)$

7 To a solution of each salt; add dilute nitric acid and then silver nitrate solution; a white precipitate of AgCl forms with lithium chloride; and a yellow precipitate of AgI forms with lithium iodide.
$Ag^+(aq) + Cl^-(aq) \rightarrow AgCl(s)$
$Ag^+(aq) + I^-(aq) \rightarrow AgI(s)$

8 (a) chlorine: purifying drinking water; sodium chlorate(I): bleach
(b) water and chlorine gas

Chapter 10

1 Displayed formula:

```
   H  H  H  H  H  H
   |  |  |  |  |  |
H—C—C—C—C—C—C—H
   |  |  |  |  |  |
   H  H  H  H  H  H
```

structural formula: $CH_3(CH_2)_4CH_3$
molecular formula: C_6H_{14}
empirical formula, C_3H_7

2 (a) *n*-pentane,
(b) 2-methylbutane
(c) 2,2-dimethylpropane

3 $C_9H_{20}(g) \rightarrow C_2H_4(g) + C_7H_{16}(g)$

4 $350\,cm^3$

5 $CH_4(g) + \frac{3}{2}O_2(g) \rightarrow CO(g) + 2H_2O(l)$

6 Oxides of nitrogen, carbon monoxide; both can be removed using a catalytic converter.

Chapter 11

1 (a) CFCs produce chlorine radicals in the presence of ultraviolet light; these then attack ozone molecules.
(b) A particle (atom, molecule or ion); with one or more unpaired electrons.
(c) $Cl\bullet + O_3 \rightarrow ClO\bullet + O_2$; then $ClO\bullet + O_3 \rightarrow 2O_2 + Cl\bullet$

2 (a) $C_2H_6(g) + Br_2(g) \rightarrow C_2H_5Br(l) + HBr(g)$
(b) Free-radical substitution
(c) Initiation: $Br_2 \rightarrow 2Br\bullet$
Propagation: bromine radicals react with ethane molecules to form new radicals and molecules:
$C_2H_6 + Br\bullet \rightarrow HBr + \bullet C_2H_5$ then: $\bullet C_2H_5 + Br_2 \rightarrow C_2H_5Br + Br\bullet$
Termination: radicals react with each other to form molecules:
$\bullet C_2H_5 + Br\bullet \rightarrow C_2H_5Br$ or $\bullet C_2H_5 + \bullet C_2H_5 \rightarrow C_4H_{10}$

Chapter 12

1

Z-pent-2-ene   E-pent-2-ene

2 As written, from left to right, *E* (*trans*), *Z* (*cis*), *E* (*trans*), *Z* (*cis*), *Z* (*cis*), *E* (*trans*).

3 (a) $(CH_3)_2C=CH_2$

(b) $H_3C-\overset{+}{C}(CH_3)_2$ or $H_2\overset{+}{C}-CH(CH_3)_2$

(c) The tertiary carbocation:

$H_3C-\overset{+}{C}(CH_3)-CH_3$

(d) $H_3C-CBr(CH_3)-CH_3$

4 (a) $CH_3-CHBr-CH_3$

(b) $CH_2Br-CHBr-CH_3$

(c) $CH_3-CH(OSO_2OH)-CH_3$

(d) $CH_3-CH(OH)-CH_3$

5 $-C(CH_3)H-CH_2-C(CH_3)H-CH_2-C(CH_3)H-CH_2-$

Chapter 13

1 $CH_3CH_2CH_2CH_2CH_2-O-H$

2 Pentanal   Pentanoic acid

3 Butanone   Butanal

4 Silver nitrate and ammonia solution (Tollens' reagent); this is heated separately with both substances; butanal will form a silver mirror; butanone will not.

Chapter 14

1 propanoic acid and propan-2-ol

2 Add sodium carbonate, Na_2CO_3, to each sample; propanoic acid would fizz and produce carbon dioxide, which can be positively tested using limewater; propan-2-ol results in no reaction.

or

Add acidified potassium dichromate(VI) solution to each compound and warm gently; propan-2-ol will result in an orange solution turning green; no reaction takes place with propanoic acid.

3 $CH_3-CH_2-COO^-\ Na^+$

Sodium propanoate

or

$CH_3-CO-CH_3$

Propanone

4 $CH_3CH_2CH_2-O-H$

5 C–C; C–H; C–O; O–H

6 likely wavenumbers: 3000–3300 cm^{-1} (broad) for the OH; 1000–1300 cm^{-1} for the C–O

7 Likely to be due to a C=O bond; the sample therefore contains an impurity such as propanal or propanoic acid; formed by the oxidation of propan-1-ol.

Chapter 15

1 Polarising cations that have a high charge density induce more covalent character in their compounds by polarising or distorting the spherical anion. Therefore both calcium compounds are likely to be more covalent; particularly if combined with a large negative ion like the sulfide ion, S^{2-}. So, the correct answer is **A**.

2 Lattice enthalpy depends on the same factors as hydration energy — ionic charge and ionic radius. The substance with the most exothermic lattice formation energy will be the one consisting of small, highly charged positive ions (a high charge density) and also negative ions that have a high charge and small size. The correct answer is **B**.

3 Charge and size govern how strongly ions are attracted to water molecules — a high charge and small ionic radius are favourable for strong attraction. Three of them have a single charge; the calcium ion is the only one with a double charge. Hydration energy is considerably more affected by charge than ionic radius, so **B** will be the most exothermic.

4 (a) $\Delta H = 944 + (2 \times +436) - [145 + (4 \times +388)]$
$= 1816 - 1697 = +119\,kJ\,mol^{-1}$

(b) $\Delta H = 157 + (2 \times +463) - (436 + 496) = 1083 - 932$
$= +151\,kJ\,mol^{-1}$

(c) $\Delta H = 157 + (2 \times +463) - [(2 \times +463) + \frac{1}{2} \times + 496)]$
$= 1083 - 1174 = -91\,kJ\,mol^{-1}$

5 The enthalpy of formation of hydrazine assumes the formation of liquid hydrazine, not gaseous hydrazine; mean bond energies were used in calculating ΔH in Q4 part (a) rather than specific values.

6 (a) Decrease

(b) Increase

(c) Increase

(d) Decrease

(e) Decrease

7 (a) $\Delta S^{\ominus} = \Sigma S^{\ominus}_{products} - \Sigma S^{\ominus}_{reactants}$
$= +93 - [40 + 214] = 93 - 254 = -161\,J\,K^{-1}\,mol^{-1}$

(b) $CO_2(g)$ is being used up in the reaction; the number of moles of gas decreases; and more order will form in the reaction; or there will be a decrease in disorder.

Chapter 16

1 (a) Order with respect to $(CH_3)_3CBr = 1$; order with respect to $OH^- = 0$

(b) Overall order = 1 + 0 = 1

(c) $\frac{mol\,dm^{-3}\,s^{-1}}{mol\,dm^{-3}} = s^{-1}$

2 (a) Order with respect to NO = 2; order with respect to $H_2 = 1$

(b) Overall order = 2 + 1 = 3

(c) Rate = $k[NO]^2[H_2]^1$; or simplified as = $k[NO]^2[H_2]$

(d) Rate = $k[NO]^2[H_2]^1$; substituting data from experiment 1:
$1.11 \times 10^{-3} = k(0.100)^2(0.100)$; $k = 1.11\,mol^{-2}\,dm^6\,s^{-1}$

Chapter 17

1 (a) It decreases.

(b) The forward reaction is exothermic because an increase in temperature favours the reverse process – the endothermic route; K_c decreases with increasing temperature.

(c) A graph is plotted of the natural log of the equilibrium constant ($\ln K_c$) against the reciprocal of temperature (K^{-1}).

(d) (i) The gradient of the straight line is $-\Delta H°/R$, so $-R$ multiplied by the gradient is equal to the enthalpy change in $J\,mol^{-1}$.

(ii) The intercept is equal to $\Delta S°/R$, so R multiplied by the intercept gives the entropy change in $J\,K\,mol^{-1}$.

Chapter 18

1 (a) $Cr_2O_7^{2-}(aq) + 14H^+(aq) + 6e^- \rightarrow 2Cr^{3+}(aq) + 7H_2O(l)$
$2I^-(aq) \rightarrow I_2(aq) + 2e^-$

(b) (i) Chromium(VI) in dichromate(VI)

(ii) Iodide ions

(c) $Cr_2O_7^{2-}(aq) + 14H^+(aq) + 6I^-(aq) \rightarrow 2Cr^{3+}(aq) + 3I_2(aq) + 7H_2O(l)$

(d) (i) +6 to +3

(ii) −1 to 0

2 (a) $FeO_4^{2-}(aq) + 8H^+(aq) + 3e^- \rightarrow Fe^{3+}(aq) + 4H_2O(l)$
$MnO_2(s) + 2H_2O(l) \rightarrow MnO_4^-(aq) + 4H^+(aq) + 3e^-$

(b) (i) Iron(VI) in ferrate(VI)

(ii) Manganese(IV) in MnO_2

(c) $FeO_4^{2-}(aq) + 4H^+(aq) + MnO_2(s) \rightarrow Fe^{3+}(aq) + 2H_2O(l) + MnO_4^-(aq)$

(d) (i) +4 to +7

(ii) +6 to +3

3 $Fe^{3+}(aq) + e^- \rightleftharpoons Fe^{2+}(aq)$; $E^{\ominus} = +0.77\,V$, so iron(III) ions will react with any halide ions associated with electrode potentials less positive; that is: $I_2(aq) + 2e^- \rightleftharpoons 2I^-(aq)$; $E^{\ominus} = +0.54\,V$; the only reaction that takes place is
$2Fe^{3+}(aq) + 2I^-(aq) \rightarrow 2Fe^{2+}(aq) + I_2(aq)$.

4 (a) (i) No

(ii) Yes

(iii) No

(iv) No

(v) Yes

(b) (i)

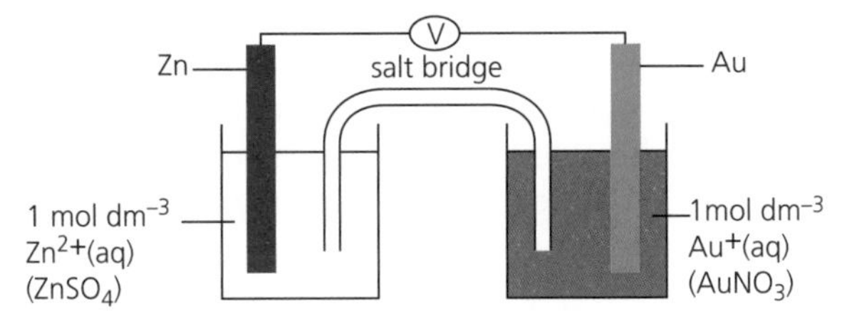

(ii) $E^\ominus_{cell}$ = +1.68 + (+0.76) = +2.44 V

(iii) Zn(s) | Zn^{2+}(aq) || Au^+(aq) | Au(s)

Chapter 19

1 HNO_3 is base 1; $H_2NO_3^+$ is conjugate acid 1.
H_2SO_4 is acid 2; HSO_4^- is conjugate base 2.

2 (a) $C_6H_5COOH(aq) + H_2O(l) \rightleftharpoons C_6H_5COO^-(aq) + H_3O^+(aq)$

(b) $CH_3NH_2(aq) + H_2O(l) \rightleftharpoons CH_3NH_3^+(aq) + OH^-(aq)$

3 (a) pH = $-\log_{10}[H_3O^+] = -\log_{10}(6.60 \times 10^{-2}) = 1.18$

(b) $K_w = [H_3O^+][OH^-(aq)]$

$$[H_3O^+] = \frac{1.00\times10^{-14}}{5.67\times10^{-4}} = 1.76\times10^{-11}\,\text{mol dm}^{-3}$$

pH = $-\log_{10}[H_3O^+] = -\log_{10}(1.76 \times 10^{-11}) = 10.75$

(c) $K_w = [H_3O^+(aq)][OH^-(aq)]$

$$[H_3O^+] = \frac{1.00\times10^{-14}}{0.0500} = 2.00\times10^{-13}\,\text{mol dm}^{-3}$$

pH = $-\log_{10}[H_3O^+] = -\log_{10}(2.00 \times 10^{-13}) = 12.70$

(d) pH = $-\log_{10}[H_3O^+] = -\log_{10}(2 \times 0.0950) = 0.72$

4 $[H_3O^+] = 10^{-pH} = 10^{-6.45} = 3.55 \times 10^{-7}$ mol dm^{-3}

$K_w = [H_3O^+(aq)][OH^-(aq)]$

$$[OH^-] = \frac{1.00\times10^{-14}}{3.55\times10^{-7}} = 2.82\times10^{-8}\,\text{mol dm}^{-3}$$

5 (a) $CH_3COOH(aq) + H_2O(l) \rightleftharpoons CH_3COO^-(aq) + H_3O^+(aq)$

(b) $$K_a = \frac{[CH_3COO^-(aq)][H_3O^+(aq)]}{[CH_3COOH(aq)]}$$

(c) $[H_3O^+] = \sqrt{K_a \times [HA]} = \sqrt{1.74\times10^{-5}\times0.001}$
$= 1.32 \times 10^{-4}$ mol dm^{-3}

pH = $-\log_{10}(1.32 \times 10^{-4}) = 3.88$

6 (a) $C_6H_5COOH(aq) + H_2O(l) \rightleftharpoons C_6H_5COO^-(aq) + H_3O^+(aq)$

or:

$C_6H_5COOH(aq) \rightleftharpoons C_6H_5COO^-(aq) + H^+(aq)$

(b) The equilibrium constant for the dissociation is given by:

$$K_a = \frac{[C_6H_5COO^-(aq)][H^+(aq)]}{[C_6H_5COOH(aq)]}$$

$pK_a = -\log_{10}K_a$, so K_a = antilog$_{10}$ (−4.20) = 6.31×10^{-5} mol dm^{-3}

pH = $-\log_{10}[H^+(aq)]$ so H^+ = antilog$_{10}$ (−4.50) = 3.16×10^{-5} mol dm^{-3}

Substituting gives:

$$6.31 \times 10^{-5} = \frac{[C_6H_5COO^-(aq)]\times3.16\times10^{-5}}{[C_6H_5COOH(aq)]}$$

$$\frac{[C_6H_5COO^-(aq)]}{[C_6H_5COOH(aq)]} = \frac{6.31\times10^{-5}}{3.16\times10^{-5}} = 2$$

Therefore

$$\frac{[C_6H_5COOH(aq)]}{[C_6H_5COO^-(aq)]} = \frac{1}{2} = 0.5$$

Or:

(c) (i) pK_a would stay the same

(ii) The dissociation of an acid is exothermic, and so an increase in temperature would favour the endothermic process – the reverse process. The concentrations of hydrogen ions and benzoate ions would therefore decrease and K_a would decrease. pK_a would increase because $pK_a = -\log_{10} K_a$, so as K_a decreases, pK_a increases in value.

Chapter 20

1 (a) $4Al(s) + 3O_2(g) \rightarrow 2Al_2O_3(s)$

(b) $S(s) + O_2(g) \rightarrow SO_2(g)$

(c) $4Na(s) + O_2(g) \rightarrow 2Na_2O(s)$

2 $2Cs(s) + 2H_2O(l) \rightarrow 2CsOH(aq) + H_2(g)$; the likely pH of the solution formed will be 13–14; caesium hydroxide is a strong alkali.

3 (a) $SO_2(g) + H_2O(l) \rightarrow H_2SO_3(aq)$

(b) $Na_2O(s) + H_2O(l) \rightarrow 2NaOH(aq)$

4 $2NaOH(aq) + H_2SO_3(aq) \rightarrow Na_2SO_3(aq) + 2H_2O(l)$

Chapter 21

1 (a) 4

(b) 6

(c) 4

(d) 6

(e) 6

2 (a) Ethanedioate ions

(b) Octahedral

(c) +3

(d) $[Ni(C_2O_4)_3]^{4-}$

3 (a) 2

(b) $[Fe(ind)_3]^{2+}$

4 $Cr_2O_7^{2-} + 14H^+ + 6I^- \rightarrow 3I_2 + 2Cr^{3+} + 7H_2O$

moles of dichromate(VI) = $\frac{21.25}{1000} \times 0.0100$
$= 2.125 \times 10^{-4}$ mol

moles of iodide = $2.125 \times 10^{-4} \times 6 = 1.275 \times 10^{-3}$ mol

total moles of iodide = $1.275 \times 10^{-3} \times \frac{250}{25} = 0.01275$ mol mass of iodide = 0.01275 × 126.9 = 1.618 g

So:

percentage by mass of iodine as I^- = $\frac{1.618}{50.0} \times 100 = 3.24\%$

5 (a) A pale blue precipitate forms:

$[Cu(H_2O)_6]^{2+}(aq) + 2OH^-(aq) \rightarrow [Cu(H_2O)_4(OH)_2](s) + 2H_2O(l)$

or:

$Cu^{2+}(aq) + 2OH^-(aq) \rightarrow Cu(OH)_2(s)$

This dissolves in excess ammonia to form a dark blue solution:

$[Cu(H_2O)_4(OH)_2](s) + 4NH_3(aq) \rightarrow [Cu(NH_3)_4(H_2O)_2]^{2+}(aq) + 2H_2O(l) + 2OH^-(aq)$

(b) A dark green precipitate forms:

$[Cr(H_2O)_6]^{3+}(aq) + 3OH^-(aq) \rightarrow [Cr(H_2O)_3(OH)_3](s) + 3H_2O(l)$

or:

$Cr^{3+}(aq) + 3OH^-(aq) \rightarrow Cr(OH)_3(s)$

This dissolves in excess NaOH to form a green solution:

$[Cr(OH)_3(H_2O)_3](s) + 3OH^-(aq) \rightarrow [Cr(OH)_6]^{3-}(aq) + 3H_2O(l)$

(c) A white precipitate forms and fizzing takes place:

$2[Al(H_2O)_6]^{3+}(aq) + 3CO_3^{2-}(aq) \rightarrow 2[Al(H_2O)_3(OH)_3](s) + 3CO_2(g) + 3H_2O(l)$

6 Sodium carbonate will form a brown precipitate of iron(III) hydroxide with iron(III) and also fizzing takes place; with iron(II) ions, a dark green precipitate of iron(II) carbonate forms and there is no fizzing.

7 (a) 6

(b) $[Fe(H_2O)_6]^{2+}(aq) + TPEDA \rightleftharpoons [Fe(TPEDA)]^{2+}(aq) + 6H_2O(l)$

(c) Lots of water molecules are released; so entropy increases in the ligand substitution process; creating a more negative ΔG.

Chapter 22

1 (a) Molecules with the same molecular formula but different structures.

(b) $CH_3CH_2CH_2CH_2CH_2CH_3$ Hexane; $CH_3CH(CH_3)CH_2CH_2CH_3$ 2-Methylpentane; $CH_3CH_2CH(CH_3)CH_2CH_3$ 3-Methylpentane; $CH_3CH(CH_3)CH(CH_3)CH_3$ 2,3-Dimethylbutane; $(CH_3)_3CCH_2CH_3$ 2,2-Dimethylbutane

2 (a) $C_{16}H_{30}O_2$

(b) It has two hydrogen atoms on each side on the carbon–carbon double bond that can be either on the same side (*Z*) of the double bond or on opposite sides (*E*); lack of free rotation about the carbon–carbon double bond makes this possible.

(c) $CH_3(CH_2)_5(H)C{=}C(H)(CH_2)_7COOH$ (H and $(CH_2)_7COOH$ on top; $CH_3(CH_2)_5$ and H below)

3 (a) Molecules having the same formula but different spatial arrangements of atoms.

(b) Two mirror-image structures: C bonded to COOH, H, OH, H_3C | C bonded to COOH, HO, H, CH_3

(c) One enantiomer will rotate the plane of polarised light in one direction; the other will rotate plane-polarised light in the other direction; to the same extent (at the same concentration).

Chapter 23

1 Aldehydes: cinnamaldehyde and vanillin; ketones: carvone and progesterone

2 (a) $C_6H_5CH_2OH$

(b) $(H_3C)_2CH{-}CH_2{-}OH$

(c) $H{-}CH_2{-}CH_2{-}CH_2{-}CH_2{-}OH$

3 $C_6H_5{-}C(HO)(CN){-}CH_2{-}C_6H_4{-}CH(HO)(CN)$

4 (a) $CH_3CH_2COO^-Na^+$ + CH_3CH_2OH

(b) 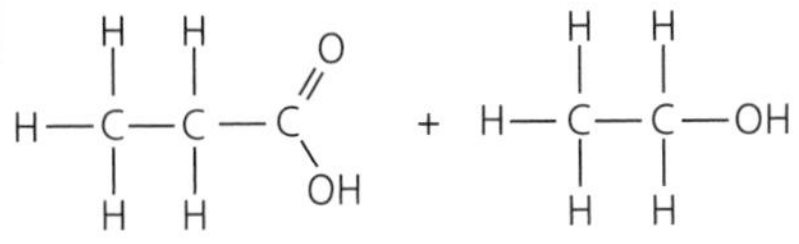 H H H–C–C–C(=O)OH H H + H H H–C–C–OH H H

5 O=C(CH$_3$)–HN–(cyclohexyl)

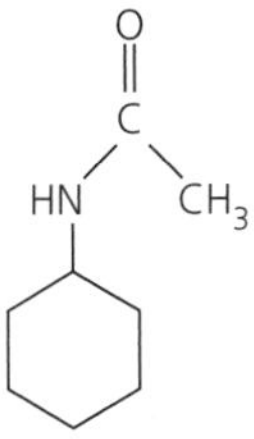

Chapter 24

1 (a) Reaction 1: N^+O_2

Reaction 2: CH_3C^+O

(b) Reaction 1: concentrated nitric(V) acid and concentrated sulfuric(VI) acid; maximum temperature 50°C

Reaction 2: CH_3COCl; $AlCl_3$ or $FeCl_3$

Chapter 25

1 (a) 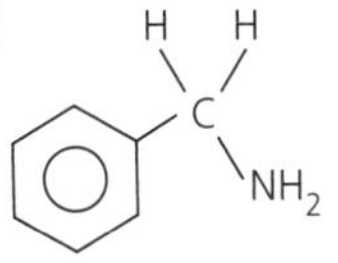 H H C NH$_2$

(b) 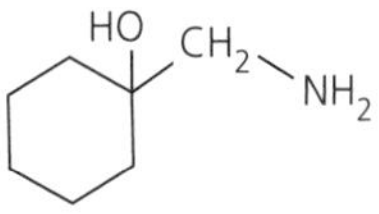 HO CH$_2$ NH$_2$

2 (a) 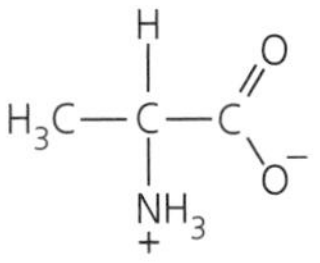 H H$_3$C–C–C(=O)O$^-$ NH$_3^+$

(b) H H$_3$C–C–C(=O)OH NH$_3^+$

(c) COOH CH$_3$ CH$_2$ H$_2$N–C–C–N–C–COOH H O H H

COOH CH$_2$ CH$_3$ H$_2$N–C–C–N–C–COOH H O H H

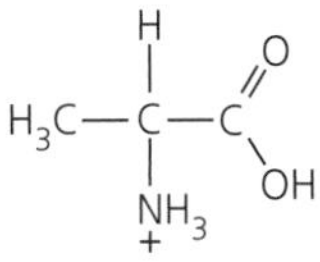

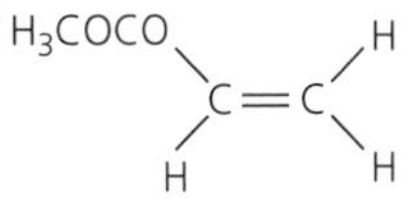

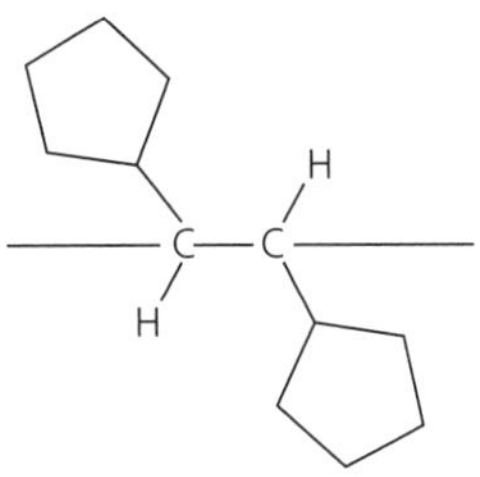

CH$_3$ CH$_3$ H$_2$N–C–CH$_2$–C–N–C–COOH H O H H

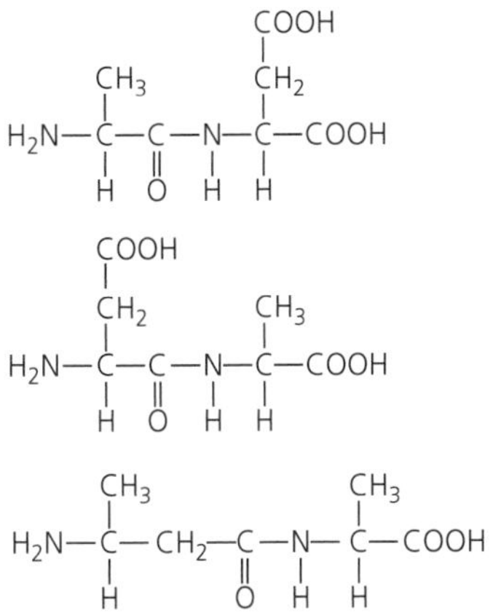

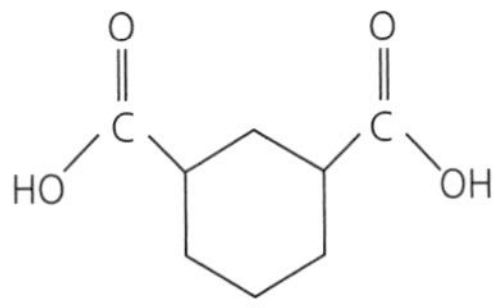

Chapter 26

1 (a) 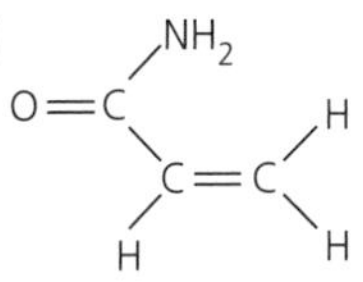 NH$_2$ O=C H C=C H H

(b) H$_3$COCO H C=C H H

2 H C–C H

3 O O C C HO OH and CH$_2$ CH$_2$ HO OH

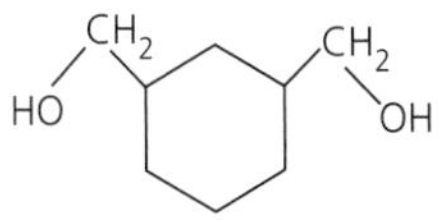

4 (a) An amino acid

(b) 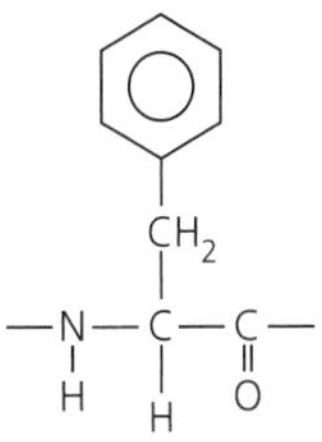 CH$_2$ –N–C–C– H H O

(c) CH$_2$ CH HN C=O O=C NH CH CH$_2$

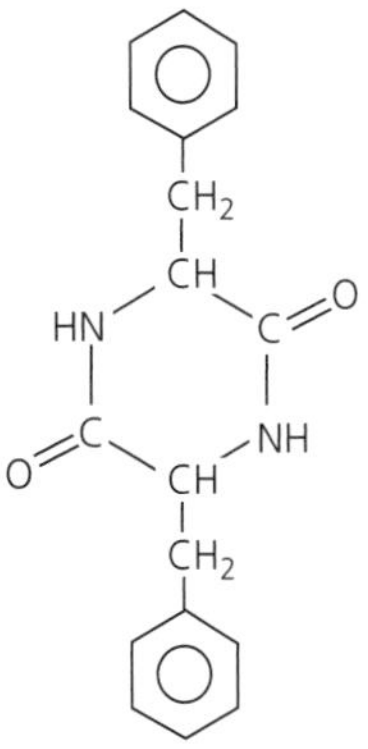

Chapter 27

1 (a) OH; a stretch occurs at about 3350 cm^{-1}

(b) $M_r = 60$

(c) (i) 26.7% oxygen

amounts in mol: $\frac{60}{12.0}$ = 5.0 mol carbon

$\frac{13.3}{1.0}$ = 13.3 mol hydrogen

$\frac{26.7}{16.0}$ = 1.67 mol oxygen

The simplest ratio is 3 : 8 : 1 or C_3H_8O.

(ii) C_3H_8O

(d) X must have an OH present and contain three carbon atoms; so could be either propan-1-ol or propan-2-ol:

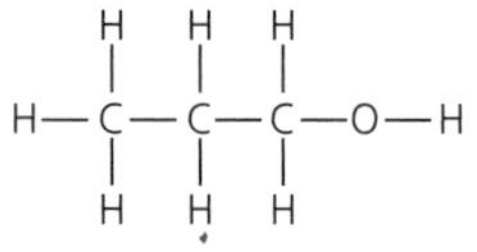

```
    H   H   H
    |   |   |
H—C—C—C—O—H
    |   |   |
    H   H   H
```

Propan-1-ol

```
    H   H   H
    |   |   |
H—C—C—C—H
    |   |   |
    H   O   H
        |
        H
```

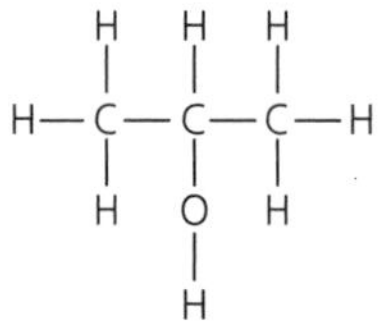

Propan-2-ol

2 The molecule has three proton environments that will generate areas in the ratio 1 : 2 : 1 (moving from left to right along the molecule). The left-hand CHO group will give a peak at δ 10–11; the right-hand CHO group will give a peak at a slightly different δ value; the central CH_2 group would give a peak at a higher δ of 5–6.

The peak from the CHO group on the left would be a singlet. The CH_2 protons peak would be split into a doublet by the proton in the right-hand CHO group. The peak from the CHO group on the right would be split into a triplet by the protons in the CH_2 group.